Praise for *Floored!*

"One of the best slip and fall experts I have ever used. A true professional!"
—**Debra K. Cook**, Attorney at Law, Cannon & Nelms, A Professional Corporation

"We retained Russell J. Kendzior in a floor mat trip and fall case, and after a long and difficult battle we were able to overturn a lower courts summary judgment at the state supreme court, which lead to a $3.95 million settlement."
—**Scott Blumenshine**, Meyer and Blumenshine, Chicago, Illinois

"Kendzior is one of the most knowledgeable, passionate, and focused expert witnesses I have had the pleasure to work with. A true professional."
—**William K. Goldfarb**, the Law Offices of William K. Goldfarb

"Kendzior is the expert's expert!"
—**Charles Dunkel**, Shirley Mills, LLP, Houston, Texas

"Kendzior's trial presence is without a doubt one of the best we have seen of any expert witness we have hired."
—**Terry Evans**, Andereck, Evans, Widger, Lewis & Figg, LLC

"Kendzior is the Albert Einstein of the walkway safety industry."
—**Benjamin Hall**, Hall Law Firm, Houston, Texas

FLOORED!

FLOORED!

Real-Life Stories from a Slip and Fall Expert Witness

Russell J. Kendzior

ROWMAN & LITTLEFIELD
Lanham • Boulder • New York • London

Published by Rowman & Littlefield
A wholly owned subsidiary of The Rowman & Littlefield Publishing Group,
Inc.
4501 Forbes Boulevard, Suite 200, Lanham, Maryland 20706
www.rowman.com

Unit A, Whitacre Mews, 26-34 Stannary Street, London SE11 4AB

British Library Cataloguing in Publication Information Available

Library of Congress Cataloging-in-Publication Data

Names: Kendzior, Russell J., author.
Title: Floored! : real-life stories from a slip and fall expert witness / Russell J. Kendzior.
Description: Lanham : Rowman & Littlefield, [2017]
Identifiers: LCCN 2017018422| ISBN 9781442271692 (cloth : alk. paper) | ISBN 9781442271708
(ebook)
Subjects: LCSH: Falls (Accidents)
Classification: LCC RD93 .K47 2017 | DDC 617.1—dc23 LC record available at https://
lccn.loc.gov/2017018422

∞ ™ The paper used in this publication meets the minimum requirements of American National
Standard for Information Sciences Permanence of Paper for Printed Library Materials, ANSI/
NISO Z39.48-1992.

Printed in the United States of America

This book is dedicated to the blue-collar men who served as my role models, helped shape my life, and taught me right from wrong. My father, Joe; my uncle, Ray; and my father-in-law, John. Three men from the greatest generation.

CONTENTS

FOREWORD

Every time I am asked, "So what do you do?" I reply, "I am a safety expert and expert witness in cases involving a slip, trip, and fall," to which they usually reply, "Really? Slips and falls? Aren't those just scams staged by con artists?" Welcome to my world. A world of misunderstanding and mischaracterization of slip, trip, and fall victims. I don't think there is any category of safety whose victims are demonized more than that of slips, trips, and falls.

The question has now turned into a conversation where I am asked to describe a typical case that I have worked on. "Well, there was a case where while attending her husband's funeral, a woman slipped and fell, striking her head on her husband's casket (chapter 3) or the case where a young boy tripped and fell while playing laser tag and contracted a flesh-eating bacteria (chapter 2)," to which they usually respond by saying, "Wow! I never realized people can really get hurt by a slip and fall."

For the past twenty-plus years I have been taking cases as an expert witness. It has been my experience that very few slip, trip, and fall lawsuits are filed by fraudulent scam artists seeking to make a fast buck but are rather just regular folks like you who get seriously hurt and usually just want to get their medical bills paid.

Much of the distortion is generated by the television media, who sensationalizes fraudulent scam artists who are often caught on surveillance video but rarely will broadcast a story on legitimate slips, trips, and falls. It's estimated that approximately 3 percent of all slip, trip, and fall claims are fraudulent, leaving the remaining 97 percent as legitimate. Therefore, for every three slip-and-fall con artists caught on store surveillance video there are 97 legitimate slips and falls recorded. Sadly, they don't get any airtime. This distortion of reality is in part what this book seeks to correct.

To suggest that all slip, trip, and fall lawsuits are frivolous is untrue, and to suggest that people should simply watch where they're walking is a gross overexaggeration of the underlying and serious problem. Denying that wet floors contribute to slips and falls is like saying that cigarette use does not contribute to lung cancer. Like the tobacco industry, who for decades denied the link between cigarette use and cancer and said that smokers can stop smoking anytime they want, the floor covering industry has adopted the same playbook of denying that their products contribute to slips and falls and usually blame the victim for their fall. Questions like "What type of shoes were you wearing?" and "Where were you looking at the time you fell?" are commonly asked of plaintiffs for the purpose of transferring responsibility and liability. Floor covering manufacturers will also blame their commercial customers by asking, "What floor cleaner do you use?" therefore suggesting that everyone is to blame except the flooring manufacturer. The cost of slips, trips, and falls is huge, topping an estimated $100 billion each year, and passing the buck hasn't helped. The problem isn't just floor covering manufacturers but property owners who simply don't care. When asked, "Do you have many slip-and-fall claims?" many business owners will respond by saying, "Sure, that's why we have insurance." They see slip-and-fall claims like they do their electric bill. A cost of doing business that they have little control over and in turn pass along to their customers via the products they sell.

I once asked a grocery store executive what their slip-and-fall prevention strategy was, whereby he replied, "We simply raise the cost of bananas and blame the problem of 'frivolous lawsuits' on the trial attorneys" (chapter 43). To him the problem was lawyers filing lawsuits and not his stores' ongoing hazardous conditions that lead to his customers becoming injured. This mindset is in many ways the real problem and explains why slip, trip, and fall injuries and lawsuits continue to rise.

Since my first retention in 1997 I have worked on more than 750 slip, trip, and fall lawsuits, representing both plaintiffs and defendants, and receive calls almost every day from attorneys across the country who are seeking to retain an expert witness. Sadly, in my line of work there are no winners, just losers. For plaintiffs who are often seriously injured there is no sum of money that can make them whole again, and for the defendant the rising costs of litigation can break their bottom line and dramatically increase their insurance rates.

Rarely have I been presented a case of fraud, and for good reason. Most personal injury plaintiff attorneys will take cases on a contingency basis whereby they must front the cost of litigation and don't want to waste their time and money on a weak case. Slip, trip, and fall lawsuits are hard for plaintiffs to win; just ask any plaintiff attorney. That's not to say that they are not winnable, just that the burden is high and most attorneys simply do not want to take the financial risk.

Most of the lawsuits I have been retained in settle out of court, leaving just a small percentage that go to trial. I have given hundreds of depositions and testified at dozens of trials. I have worked in cases from coast to coast and internationally. You never know how the case will end. I once worked on a case where the defendant offered to settle the matter for $5,000, which went to court and the plaintiff won a $600,000 jury verdict. I have also worked on a case where a plaintiff was offered nearly a million dollars to settle their case, went to court, and lost.

Since the dawn of time people have enjoyed a good story, which I have many to tell. This book examines just a small fraction of the cases I have worked on during my twenty-year career as an expert witness, which I hope you find both fascinating and informative. There is no levity in describing people's pain and suffering, but I hope that this book can provide a bit of insight into real-world slip, trip, and fall lawsuits and the people involved.

I am a blue-collar man working in a white-collar profession. I was raised to value hard work and honesty and to have a deep sense of fairness and justice for all people. I am not a plaintiff or defense advocate but rather take cases based on their merit. If I can help, I will; if I can't, I don't. It's that simple to me. It's difficult sometimes to not carry or forget the pain that I often see in the faces of those who have been injured. These are real people, and their stories need to be told.

The cases I discuss are based on real lawsuits; however, some of the information, including names and locations, has been modified so as to protect the privacy of the parties involved.

I

GROCERY CROOKS!

On a peaceful March afternoon in Houston, Texas, Ms. Garcia, accompanied by her daughter-in-law and granddaughter, went to her local grocery store. Ms. Garcia was walking next to her daughter-in-law, and trailing them was their young granddaughter pushing a shopping cart. As they approached the dairy aisle, Ms. Garcia suddenly and unexpectedly slipped and fell, injuring her right hip and leg. Her daughter-in-law quickly summoned a store employee by the name of Miller to come to their aid. Upon arrival Miller quickly called his manager, Mr. Crooks, who arrived just moments later. Miller left to get a chair for Ms. Garcia while Crooks began to take down information for an incident report. Upon arrival Miller was asked to clean up a broken egg that was smeared on the floor next to Ms. Garcia. Miller later stated in his deposition that he cleaned up two eggs separated by approximately two to three feet. Sadly Ms. Garcia was too severely injured to sit in a chair, and Crooks called 911, who appeared at the store just minutes later. Ms. Garcia was transported to the local hospital and diagnosed with a broken hip and leg. Although Mr. Crooks was very attentive at the time of Ms. Garcia's injury, once she was transported from his store, he made no attempt to contact her to see how she was doing.

Pursuant to the suit the defendant produced the incident report as prepared by Crooks. Contained in the incident report were a series of small errors that collectively raised questions about its accuracy. For example, although Crooks attended to Ms. Garcia just moments after her fall, he reported the time of her event incorrectly, claiming it had occurred more than an hour later. Crooks would state in his deposition that he had not spoken to Miller about Ms. Garcia's fall, but in Miller's witness statement, which was attached to the incident report, Miller said that when asked for the time of the fall by Crooks, Miller said it was 4:21 and not 3:17, as can be seen on the store's surveillance video that captured Ms. Garcia. Another seemingly minor inaccuracy was the mis-identification of Ms. Garcia's shoes. According to Crooks, he recorded that Ms. Garcia was wearing a pair of slip-on shoes, when in fact she was wearing a pair of Dr. Scholl's shoes. During his deposition Crooks said that he never spoke to Miller, which was clearly incorrect based on Miller's written statement. Crooks also said that he did not actually see an egg on the floor and that he did not perform an inspection of the floor because his primary concern was for the care of Ms. Garcia. Mill-er, however, stated in his deposition that shortly after Crooks arrived, he was told to "clean up the egg," which he can be seen doing on store surveillance. What is also captured on surveillance video is Miller stock-ing eggs in the display approximately thirty minutes prior to Ms. Gar-cia's slip and fall. Miller claims in his signed statement that he did not see an egg on the floor, but store surveillance video shows him kicking an object on the floor as he approached the refrigerated egg display.

What was not seen on the one-hour-and-seven-minute video was anyone dropping an egg. The only time any type of image was captured on store surveillance video was just a few minutes before Ms. Garcia's slip and fall.

THE TRIAL

At trial, I was asked what my opinions were as related to how and when the egg was dropped. I responded that there is no evidence of anyone dropping an egg at the location where Ms. Garcia later slipped and fell. So by reason of deduction the egg must have been dropped prior to the start of the surveillance video. After a lengthy sidebar the judge dis-missed the jury to discuss if my opinions were to be accepted by the court given that I was not able to identify when the egg fell and who dropped it. The judge finally ruled that given that neither I nor defense counsel could state with certainty exactly how, when, and who dropped the egg, my opinion that suggested that the egg was dropped before the start of the video was plausible. What everyone agreed to was the fact that there was an egg on the floor at the location where Ms. Garcia slipped and fell; we just didn't know how it got there.

Defense counsel then began to question my experience, training, and knowledge as it relates to eggs and asked, "Mr. Kendzior, you claim to be an expert in walkway safety; what does that have to do with eggs?" to which I responded, "Nothing." He then began to ask if I had any experience in the retail packaging of eggs, the stocking of eggs, or the handling of eggs. He then asked the ultimate question: "Mr. Kendzior, you do not make yourself out as an egg expert, do you?" I paused for a short time and answered, "Yes, I am. You see, I raise chickens and ducks as a hobby, whereby my girls, as I call them, produce a dozen or so eggs each day. My wife and I clean them, place them in cartons (pink cartons), refrigerate them, and deliver them to a local farmer's market for sale. I am indeed an egg expert." The court was eerily silent, and the defense attorney was clearly shaken from his list of prepared questions. In fact he was left speechless for what felt like an eternity.

My opinions were straightforward and simple. Although the store had good policies, their store manager did not. He not only got the small things wrong, he missed the big things as well. Things like training his employees in slip-and-fall prevention. When deposed, Miller said he did not have any safety training; however, company policy required such. In fact, when asked how often his employees talk about safety issues, Crooks said that they "huddle" three times a day, a fact Miller again refuted. Miller said that even though it was store policy to inspect the floors for any hazardous condition, he did not and assumed someone else did. As stated earlier, Miller said in his written statement that he did not see an egg on the floor prior to Ms. Garcia's fall, but after her fall he said that he cleaned up two eggs.

In the depositions of Ms. Garcia and her daughter-in-law they both stated that they only saw the raw egg (whites and yolk) after the fall and that they did not see any eggshells in the area where the egg was. During the trial, I was asked how is it that an egg was dropped but there was no shell? I referred to the portion of the video that showed Miller approaching the egg display just minutes before Ms. Garcia's slip and fall, where he can be seen kicking something toward the display. What he was kicking was the eggshell! So when he said he cleaned up two eggs, he actually cleaned up the egg white and yolk and the shell that he had kicked under the display.

HOW DID IT END?

Given that the egg was on the floor for over an hour, the defendant failed to perform timely and proper inspections, which were not only required in their written safety policies but are customary in the retail industry. Secondly, when Miller kicked the eggshell under the display, he therefore was aware of the presence of the slip hazard and failed to properly remove it or post any warning. Although the quality of the store surveillance video was poor, shortly before Ms. Garcia slipped and fell another customer had rolled their cart through the egg, making it visible to the camera. The egg was virtually invisible to both the camera and store customers. Because of the combination of the high-gloss appearance of the bright, white-colored tiled floor, glare from overhead fluorescent lighting, and the transparent color of the egg white, the egg was an accident ready to happen. The hazard was further heightened by the amount of time it was on the floor. It was just a matter of time that someone would step on it and slip. That someone was Ms. Garcia.

The jury rendered a defense verdict, which was based in part on the fact that the person seen kicking the eggshell may not have been an employee but another guest. I hope that one day I get retained in a case involving a slip and fall in the honey aisle. I am also a beekeeper!

2

TAG, YOU'RE IT!

It was a quiet Saturday afternoon in a small town in Tennessee, when ten-year-old Johnathan was dropped off by his dad at the local video arcade and laser tag amusement center. Johnathan's dad, John, and the owner of the amusement center, Bob, were friends, and all the kids in this small town met at the center to socialize. The center was abuzz with the sounds of video games, children's laughter, and a steady stream of zapping sounds coming from the laser tag area. Upon arrival Johnathan met up with his friends for an afternoon of laser tag. With the help of a handyman friend, Bob had built the laser tag arcade himself. The plywood walls, floors, and ramps created an interesting maze for those playing the game. Graffiti was painted on the walls, floors, and ramps with fluorescent paint by Bob's younger son, and when illuminated by the overhead fluorescent black lights, the structure came to life.

Johnathan suited up, entered the maze, and began playing the game. The game is played by shooting others with a backpack-mounted laser gun. Each player has a series of reflective targets on their laser suit, which when shot with the laser gun will register a hit. Each laser hit is registered on the player's backpack. After a preprogrammed number of hits they are eliminated from the game.

Within a matter of minutes Johnathan took his first hit and sought cover by running down an elevated platform that discharged onto a series of ramps. In an effort to avoid being hit again, Johnathan began running down the ramp, and because of its steep slope, he picked up speed, causing him to fall and scrape his knee on a raised nail head that secured the plywood floor. After completing the game Johnathan went to Bob and asked for a bandage for his scraped knee. Bob cleaned up the wound and placed a bandage on it, but because the scrape was deep, Bob called John and told him that Johnathan was hurt and might need a stitch or two. A short period later John came by the center to pick up Johnathan and take him to the local hospital emergency room. The treating physician said that the scrape was minor, and after a few stitches Johnathan and his dad were on their way home. Once at home Johnathan's knee started to turn red and within a matter of a few hours began to hurt. By the time Johnathan was to go to bed his knee, thigh, and leg were swollen and he began running a fever. John called 911, and a few minutes later paramedics arrived. Based on Johnathan's quickly degrading condition, the paramedics called in a Care flight helicopter to transport Johnathan to the closest trauma center, located in Memphis. The doctors of the trauma center began treating Johnathan, who was now in critical condition. Lab reports revealed that Johnathan had a flesh-eating bacteria, and it was spreading throughout his body, eventually reaching his brain. Johnathan was eventually stabilized but at a great expense. Johnathan had permanent brain damage.

THE LAWSUIT

Johnathan's family sued Bob and the corporation that owned the laser tag amusement center for negligence. Johnathan's horrific injury and the subsequent lawsuit were the talk of town and gained statewide media coverage. Also sued was the emergency doctor who treated Johnathan and the hospital he worked at. Bob's insurance company settled with Johnathan's family for an undisclosed sum, but the hospital would not. During the discovery period, it was learned that the emergency room doctor was not a licensed physician and failed to properly irrigate Johnathan's wound before closing it. Any residual bacteria would then be treated with an antibiotic. I was retained as an expert with the purpose to define whether the laser tag center, including the ramps and floors, was constructed in compliance within industry standards.

THE TRIAL

At what was expected to be a one-month-long trial I was called to testify. My opinions were that the plywood floors and ramps were improperly constructed as to not have the right fasteners. Bob and his handyman friend used flat-head nails instead of wood-fastening screws. Because people would be running across the floor, causing plywood flexing, it was likely that the nail heads would soon rise above the plywood; screws wouldn't. I also stated that the ramps, specifically the one that Johnathan was running down prior to his fall, were improperly sloped as to be too steep. The requirements as established by the building code and ADA state that ramps have a maximum slope of one in twelve, which means for every unit in height there cannot be less than twelve units in length. The ramp than Johnathan fell from had a slope of 1 in 2.5, meaning that the ramp angle was approximately twenty-five degrees. Furthermore, based on the short distance between the three adjacent ramps, anyone coming down the first ramp would build up speed as they approached the second and third ramps, making it likely that they may lose their balance and fall as they reached the plywood floor. Therefore, Bob and his handyman's design was inherently flawed and presented an unreasonably dangerous condition for the players.

At trial the hospital attorney approached the lectern and asked me if a twenty-five-degree ramp, regardless of its location, would be dangerous. My response was yes, a twenty-five-degreed sloped ramp would be dangerous. Given that the hospital's attorney asked the same question twice, I knew he was setting a trap for me. After a short pause the attorney turned around and went back to his seat, where he retrieved a poster-size image of what appeared to be a pedestrian ramp located in a building. The ramp was concrete and had handrails on either side and appeared to have a steep slope like that of the ramp at the laser tag facility. The attorney asked if I was familiar with the ramp he had blown up to poster size; I stated no. He then said that this was an image of a ramp located in the basement of the courthouse that we were in and that it too had a twenty-five-degree slope. The jury began to move in their seats. He then asked me if I would consider that the ramp in the courthouse also dangerous; I answered yes. The attorney then stated, "Really? So you're saying that the county has constructed an unsafe ramp in this courthouse?" I again affirmed my answer, and then I looked at the judge and said perhaps it would be good if the jury could go to the basement to observe the ramp for themselves. The hospital attorney broke in by saying, "That would be an excellent idea, your honor!" As he said such, the bailiff approached the judge and whispered a message to him. The judge then stated that such inspection could not take place since the basement ramp had been barricaded and was awaiting demolition and reconstruction due to complaints that it was too steep. I looked at the jury and knew the case was over. The hospital lawyer, poster board in hand, froze in amazement, returned to his chair, and said, "I have no further questions, Your Honor."

HOW DID IT END?

That day the hospital settled with Johnathan's family for an undisclosed eight-figure amount.

3

KILLER FUNERAL!

It was a cold and dreary January morning when Marie awoke and began to dress. Today was the day of her husband, Stephen's, funeral. Marie and Stephen had been married for more than forty years and raised four children. Several years ago Stephen and Marie moved from their longtime home in Indiana to Illinois, but both agreed that when the time came to "meet the Lord" they would both be laid to rest "back home" in Indiana.

The funeral home was located a few miles from Stephen and Marie's home, which was just over the Indiana state border in Illinois and just a short drive to the Indiana cemetery. It had snowed heavily the night before, and the cemetery was covered in nearly a foot of snow. The family requested a graveside service, and shortly after the final viewing Marie and her family began Stephen's final journey to Indiana. As the motorcade arrived at the cemetery, Marie noticed that the ground was covered in snow and that the cemetery attendants had not plowed a path to Stephen's grave. Stephen's family was concerned that the condition of the cemetery might not be safe and asked a cemetery worker who was at the grave if he could shovel a path. He responded, "You'll have to work that out with the funeral home people." However, there was no funeral director on the scene, only the hearse driver, who assured the family that everything would be just fine and that the funeral director was on his way. Well, wouldn't you know that the weather began to change, and a light snow began to fall. Rather than wait for the funeral director to arrive, the family decided to move on with Stephen's funeral. The family followed closely behind Stephen's casket, which was slowly being carried to the gravesite. Step by step, the pallbearers stepped through deep snow until they arrived at the grave.

As Marie approached Stephen's casket, she felt her left foot slip out beneath her, causing her to fall and strike her head on the corner of Stephen's casket. The grieving and now confused family members quickly ran to Marie, who was unresponsive and not moving. After what seemed like an hour but was only a few minutes, Marie regained consciousness. EMTs arrived at the scene, whereby she was taken to the emergency room and diagnosed with a head injury as well as a broken hip. Stephen's funeral was indeed a sad and memorable day.

THE LAWSUIT

Marie's family sued both the funeral home and cemetery for negligence, claiming that both parties acted recklessly in not providing a safe pathway for those attending Stephen's funeral. Marie's slip and fall was a result of her stepping on a granite grave marker that was covered with snow. Both codefendants pointed the finger at each other (which is common), claiming that the other party was responsible.

So who, if anyone, is to blame? My job was to establish the standard of care that funeral homes and cemeteries should follow for funerals, but this funeral was different in that there was heavy snow and no funeral director on-site. Although it is not uncommon for people to be buried on days when heavy snow has fallen, it is beyond the standard of care to not provide a safe path to and from the gravesite, and if such can not be provided, then the graveside service should be either postponed or canceled.

Also, not having a funeral director on-site is a violation of the industry standard and under Indiana law is also illegal. According to state law, a funeral director must accompany the deceased from the funeral home to the cemetery and remain with the deceased until their burial is complete. None of this had happened, and so the suit between the family and the funeral home and cemetery raged on.

For almost two years the lawyers hotly debated the facts that contributed to Marie's slip and fall. The funeral home was part of a chain owned by a gentleman who was a very successful businessman but not a particularly knowledgeable funeral director. The funeral director's grasp of the funeral business was more from the financial side and not the day-to-day operations.

The funeral director, who according to the hearse driver was on the way to the graveside service, ran that part of the business, but because he was not at the site at the time of Marie's fall, the now high-risk service went on without him. The hearse driver and cemetery worker had little to say and simply "did their job as they were told to do." The cemetery was owned and operated by a local church and took the position that either the funeral home or the family was in charge of the gravesite service and that they had little control.

The hearse driver, although technically a representative of the funeral home, was not legally responsible for the actions or inactions of the company he worked for. During his deposition the hearse driver admitted that he had no formal training other than that of driving the hearse and that he made no effort to inform the family as to their options. His statement regarding the funeral director being on the way turned out to be false. The funeral director never arrived; thus, the funeral home violated Indiana state law.

The law asks if a defendant knew or should have known of the hazardous condition prior to the event. In this case, were the funeral home and cemetery aware of the slip-and-fall risks associated with those who attended the funeral? Had the funeral director arrived on time, which he was obligated under law to do? The cemetery also had a shared responsibility in that they were aware that the funeral director was not on-site and therefore had the responsibility to delay the service until he arrived. It was my view that both codefendants were negligent in failing to provide a safe graveside service and therefore were responsible for Marie's fall and subsequent injuries.

The fact that heavy snow had fallen the night before was no surprise to the family or the two defendants, and therefore everyone was aware that Stephen's funeral might be a problem. Marie thought she did her part and came prepared with rubber snow boots. The two codefendants certainly had ample time to either rope off and plow a path to the gravesite, lay sections of plywood from the road to the grave, or at the very least inform the family as to the risk of injury and offer them the choice to cancel the gravesite funeral altogether. None of which happened. By permitting the funeral attendees to trudge through nearly knee-high snow was not a good idea and should not have taken place without the appropriate safety measures. Taking the time to scope out the cemetery grounds would seem simple, and it is. However, neither of the codefendants did so and in turn assumed the risk that someone might get hurt. The problem is, Marie did not assume the risk.

It's no surprise that the only person to actually slip and fall was Marie. She was the oldest person there. After all, one would expect to find a greater number of elderly people attending funerals, right? According to the Centers for Disease Control (CDC), falls represent the leading cause of accidental death to those over the age of eighty. Perhaps cemeteries should post a warning sign stating "Caution, cemeteries can be hazardous to your health!"

HOW DID IT END?

Three years after Marie filed her lawsuit the case settled for an undisclosed sum. What helped things along was information that suggested that the Illinois-based funeral home may not have been properly licensed to conduct funerals in Indiana, and therefore the case raised possible criminal action.

One interesting fact that arose was that Marie actually changed into rubber boots shortly before exiting the car as to ensure her safe and dry footing through the snow. But in the end it didn't matter, for it was a slippery and invisible granite grave marker beneath the snow that caused Marie to slip and fall. She could have actually died at her husband's funeral.

4

DAM WALL!

It was a cold Montana morning when Dennis went to work on the White River dam. Dennis was a good worker but was having family issues. As Dennis and his supervisor Ray drove out to the dam, which is in a remote part of northeast Montana, Ray reminded Dennis to put on his safety harness before going out onto the dam wall. The dam was originally constructed back in the 1930s and produced electricity for the county co-op. The dam consisted of a turbine generator building and a water collection area that had a narrow wall. A few years earlier a new turbine generator building was constructed to improve efficiency. The dam wall was solid concrete, and the top of the wall was narrow and uneven and was used from time to time by workers for inspection purposes. For a reason we will never know, Dennis chose not to put on his fall-restraining harness. According to his employee file, he was written up in the past for doing the same thing. Failure to wear appropriate safety equipment was a violation of company policy; according to Ray, Dennis was scheduled for termination.

Dennis was last seen by Ray walking near the dam wall without his safety harness, which was lying on the hood of Ray's service truck. Ray stated that one minute Dennis was there, then he was gone. A search-and-rescue team was dispatched, and after a three-day search found Dennis's body submerged in ten feet of water one hundred yards downstream from the dam. Dennis left a wife who was pregnant with a second child.

THE LAWSUIT

Dennis's widow sued the county, who managed the dam, and the company who built the dam, claiming that they were negligent in not providing Dennis the necessary safety equipment and that the newly constructed turbine house was improperly designed and constructed as to be unsafe.

THE TRIAL

I was retained by the construction company who designed and built the new turbine house, the only defendant in the case. The county electric co-op settled with Dennis's wife shortly after the lawsuit was filed. My role was to establish whether the turbine house and wall were safely constructed. What was puzzling was that no one actually saw how and where Dennis fell into the river. Testimony revealed that Dennis was a highly skilled swimmer who had worked for the county as a search-and-rescue scuba diver. Dennis, like all those who worked on the dam, was aware of just how dangerous this section of the White River was and that there were very strong water currents, especially around the wall of the dam. During my site inspection of the dam, Ray showed me just how strong the current was by throwing a stick into the water. The stick began to slowly sink and then spun around like it was in an underwater tornado and disappeared. Ray said, "That's why we all wear fall-restraint harnesses when we work out on the wall."

Dennis knew that his failure to wear his harness could result in tragedy if he fell into the dam but consciously chose not to wear it. The morning of his disappearance it was cold and windy. A storm was approaching, and Dennis and Ray were only there for a minor repair in the turbine house. Why Dennis chose to walk out on the wall is a mystery. What is known is that as Dennis was stepping out of Ray's truck, his wife called. Ray remembers overhearing Dennis saying to his wife that he was going to be moving out of the house. It was Ray's opinion that Dennis had intentionally jumped into the river and committed suicide, which I was very surprised to hear him say. Given that the construction company had not constructed the wall that Dennis was alleged to have fallen from, it seemed unfair that they were being sued.

My inspection of the construction design plans clearly showed that the company was not to perform any work on the dam wall, which was constructed per code.

At trial Dennis's widow was seated next to a gentleman who I later learned was her live-in boyfriend. I came to find out the child she was pregnant with at the time of Dennis's death was not her husband's but her boyfriend's. It's hard to keep secrets in this part of rural Montana. Rumors soon spread that Dennis had learned of his wife's indiscretion and was planning to divorce. Dennis's friends mentioned Dennis's concern for the welfare of his first child and that he would be seeking custody. After receiving a sizable settlement from the county co-op, Dennis's wife and boyfriend moved into a big house and were seen driving a new Harley-Davidson motorcycle. Folks in town didn't think this was right, and local folks are who serve on juries. It was Dennis's mother who provided the final piece of the puzzle by saying that her son was very depressed and distraught when he learned of his wife's cheating on him and that he probably did kill himself. She also said that her daughter-in-law told her that the money she got from the county co-op was hers and that any money she got from the construction company would go to Dennis's child.

HOW DID IT END?

The jury rendered a verdict in favor of the defendant.

5

BUSTED MY BUTT!

Kate was an attractive young twenty-eight-year-old working mom who was attending college at a state university. She worked nights and went to school during the day. One day while she was on campus, Kate made a trip to the school's administration offices. The building was nearly a hundred years old and was a historic landmark. The exterior stone facade was beautiful, and upon entering the building guests were greeted by polished limestone walls and floors, which were lighted by antique overhead lights. As Kate walked across the building's main entrance floor, she approached an arched doorway, which contained a series of three marble steps. As she stepped down the stairway, she slipped off the third step, landing on her buttocks. Kate later described the pain as like that she felt during childbirth.

Because of the severity of her injury, Kate sued the university, claiming that the steps were unsafe due to their excessive wear and lack of slip-resistant nosings.

I was retained by Kate's attorney and produced a report outlining my opinion as it related to the stairway's safety. Although the stairway was beautiful, it had shown its nearly hundred years of age. The heights between each step were very short, and the stair treads were irregularly worn; so much so that there was a trough-like area running along each tread. Such irregularities are frequently associated with pedestrian falls, especially in descent, which is how Kate slipped and fell. Finally, the edges or nosings of the treads were heavily worn and rounded smooth. It wasn't clear exactly how Kate fell down the stairs, but based on her description, it appears that she slipped off the worn tread. Also absent on the staircase were handrails. Given the age of the building and the staircase, the design elements we have today were not in place when the building was constructed. The university chose to preserve the historical characteristics of the building, which sadly were at the risk of those who used the building. Depositions of school employees revealed that there were numerous complaints about the staircase's safety, which school administrators were aware of but did not address. The case seemed pretty cut-and-dried, and I assumed that Kate's injuries were not that severe and the case would soon settle. I assumed wrong! Kate's attorney informed me that Kate was seeking more than a $1 million due to permanent disfigurement of her buttock; he then showed me photographs of her injury. I was amazed to see that upon injury, one of her butt cheeks took the brunt of her fall and had swollen to twice the size of the other cheek. The discoloration and bruising eventually subsided, but the size of her cheek was permanently much larger than the other cheek. Kate's attorney told me that according to Kate's doctor, when she fell, she broke a series of major blood vessels, which caused the muscle tissue to swell, and although it was possible they could shrink back to normal, it was also possible that her disfiguring injury might be permanent.

Without trying to be crass I had to ask, how is it that she came to claim $1 million in damages? He responded that although her medical expenses were small, her lost wages represented the bulk of her damages. Lost wages? I asked the attorney. I thought she was a college student and had a part-time job working nights. She did, he said—as a topless dancer! Her employer fired her due to her disfigured butt cheek, and based on her past five years of income, her lost wages moving forward were calculated to nearly $1 million.

HOW DID IT END?

The case settled for an amount approximately that of Kate's demand.

6

OIL SLICK!

I received a call from an attorney in Baton Rouge, Louisiana, who told me about a case he had where his fifty-four-year-old female client went into Spankey's Lube oil-change business to get her car's oil changed. A few weeks later I received the case file and began to review the case. What I read was all too familiar. The plaintiff pulled up to the front of Spankey's Lube's building and was waved into a work bay by a Spankey's Lube uniformed employee. Once stopped, one worker asked her to open her hood and began the process of changing her oil. A second employee approached her car and began entering information into a computer. Becky opened her car door, took two or three steps, and slipped and fell, striking her head on the floor. Becky was helped to her feet by the two employees and walked into the customer waiting room. A short time later one of the employees came into the waiting room, where she told him that she was not feeling well and that she had tingling in her left hand. He said, "Maybe you should not have gone out drinking last night?" Later that day Becky's tingling in her left hand became a headache, and she began to lose sensation on the left side of her body. Becky went to the emergency room and was diagnosed with a brain hemorrhage. Becky's injury eventually led to paralysis of the left side of her body.

Becky did nothing differently than millions of Americans do each year. The industry standard for the oil change business requires that only employees are to pull cars into the work bay and that customers are to stay in the waiting lounge and not enter the work bay. In fact this was Spankey's Lube's policy, which was in their employee handbook. Although the workers were aware of the policy, they often ignored it and not only permitted guests in the work bay but invited them in. Store employees were deposed and openly admitted that they frequently ignored the company's safety policies and escorted guests into the work bay.

The floors at Spankey's Lube are heavily contaminated with a wide range of automotive lubricants, which customers should not be exposed to. Spankey's Lube employees are used to the oily conditions and wear work shoes, which have slip-resistant soles. Becky was unfamiliar with the condition of the work bay and was wearing a pair of sandals.

The attorney asked that I put together a brief outline that he could present to Spankey's Lube's insurance adjuster, which I did. As a part of my research, I found that Spankey's Lube was a repeat offender and had been cited by OSHA for its slippery floors and forced to pay a $52,000 penalty.

HOW DID IT END?

The case was settled for $6 million.

7

GOT IT MAID!

It was hot and humid day when my flight landed in Monroe, Louisiana. Upon arrival I jumped into my rental car and headed south to the Winn Parish courthouse, where I was about to testify in a lawsuit involving a homeowner who hired a housekeeper to clean his house once a week. The homeowner and his wife employed the housekeeper for many years and paid her through their small business. One day as the housekeeper was about to leave, she slipped and fell on a wet floor she had just recently mopped. The homeowners were very concerned and called 911 to their home. The emergency room doctor diagnosed the housekeeper with a fractured knee, which required surgery to repair. The homeowners assured the housekeeper that she should get the necessary surgery and that they had homeowner's insurance, which should cover the cost. The homeowners contacted their insurance company and opened a claim file. Shortly after her surgery the housekeeper received a bill from the doctor and hospital and forwarded it on to her employers for submission to their insurance company. Upon receipt of the medical bills the insurance company denied the claim, stating that because the housekeeper was employed and paid by the homeowner's business, she was therefore a contractor and not a guest. Their homeowner's policy (like most policies) did not cover injuries to a person contracted to work in a person's home.

THE LAWSUIT

The suit turned out to be one of the most fascinating of my career. I was hired by the plaintiff's attorney to simply testify at trial as to a July 2003 press release published by the National Floor Safety Institute (NFSI) entitled "How the Household Cleaners Stack Up." The article addressed the drop in slip resistance as measured after a single mopping of a floor using one of the twenty household floor-cleaning agents. The plaintiff's case revolved around the NFSI's research, which showed that the floor cleaning agent she used to mop the homeowner's floor reduced the slip resistance of the floor, and therefore they provided her a dangerous substance.

THE TRIAL

Upon arriving at trial, which was a long drive in the country to the courthouse, I found the courtroom filled. I asked the bailiff who all the people were; he responded, friends and family of the homeowners. This is small-town America, and this lawsuit made its way through town. It seemed like the whole town showed up to be part of the trial. My testimony was brief and focused primarily on the household floor cleaner study. The plaintiff's attorney was a short and thin man who was wearing a white and blue seersucker suit. I thought for a moment that I was transported back in time and was in Mayberry, North Carolina, and that Andy Griffith was about to appear. Both attorneys were very polite, and my testimony was brief. For the sake of me I could not understand what the plaintiff's case was foundational on. After all, she worked for the homeowners for many years and cleaned their floors like she had done hundreds of times. Although the floor cleaner she used to mop the floor rated poorly in the NFSI test, she was aware that the floor was wet, given that she had just mopped it.

FOR IMMEDIATE RELEASE:

HOW SAFE IS YOUR FLOOR CLEANER?

Southlake, TX— November 17, 2003—Many household cleaning products leave floors
more slippery than they were before cleaning, the National Floor Safety Institute (NFSI)
reported this week.

NFSI released results of a study that showed 10 out of 19 of the best-selling household
floor cleaners on the market tested made floors more slippery than prior to use. "What
we found was shocking. Many of the household floor cleaners actually contribute to the
floor's slippery condition. Over half of the products left a slippery residue after only one
application," said Russell Kendzior, founder and executive director of NFSI. The
average product reduced the floor's slip-resistance (amount of traction) by 10%. After a
month's worth of applications, 13 out of 19 products left a slippery residue, with an
average decrease in slip-resistance of 18%.

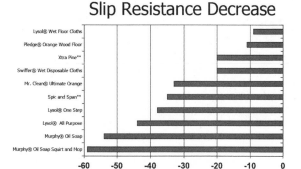

Slip Resistance Decrease

"Most of the products tested contained a strong fragrance. It was like putting perfume on
the floor. Although they made the room smell good, such fragrances are most likely the
culprit behind the slippery film," Kendzior added.

According to industry data, each year, more than one million people seek emergency
room treatment for an accidental slip-and-fall. Most of these take place in the home.

HOW DID IT END?

I learned the following week that the plaintiff prevailed and received a
judgment of $300,000. Because the homeowners did not have the funds
to pay the plaintiff, in lieu of payment she accepted their house.

8

DOUBLE TROUBLE!

It was a warm Saturday morning in Florida when Ellen went to her neighborhood food market to pick up food for a dinner party she was hosting that night. It had lightly rained that day, which is common for south Florida. As Ellen entered the market, she immediately walked to her right to retrieve a shopping cart. As she approached the carts, her feet went out from beneath her and she fell to the ground, injuring her left arm and shoulder. While on the ground, Ellen had felt that the floor was moist. It is again common in high-humidity climates for moisture to condense on a smooth, cool surface, causing the floor to "sweat." This phenomenon is like when moisture condenses on the outside of a glass of ice water.

Interestingly, floor sweating is most prominent in entryway vestibules, where humidity is higher; it is reduced deeper into the building, where air-conditioning units can more efficiently remove humidity from the air. Although there were no eyewitnesses to Ellen's fall, it was captured on the store's surveillance video, which I received and reviewed. I was retained by Ellen's attorney to provide an expert opinion as to the causation of Ellen's slip-and-fall event and whether the defendant exercised reasonable care in preventing guest-related slips and falls.

In the state of Florida, it is not required for experts to prepare written reports, but rather experts are offered to opposing counsel for deposition. A few months after retention I was deposed by the market's attorney, who spent a great deal of time addressing the facts of the case and asked, how is it that I can state with certainty that the floor was wet? And what is the definition of a wet floor? His questions reminded me of former President Bill Clinton's deposition in the Monica Lewinsky case, where he stated that "it depends on what the word is, is." The fact is, any amount of detectable moisture constitutes a wet floor, which includes floor sweating. A floor does not have to be flooded to be wet; in fact, the more water that is on the floor, the more likely it is to be seen by pedestrians than a light mist.

Given the problem of floor sweating, it was my opinion that the defendant simply installed the wrong type of entranceway floor. The solution to preventing sweating in building vestibules is to select an appropriate flooring material, preferably carpet walk-off tile. In fact, the market did have carpet runners at the entry, which worked well where they were installed, but you can't place carpet mats everywhere. The defense attorney then asked, how is it that the defendant would have known that the floors in their entranceway sweated, and did I have any evidence of sweating? Although the store manager completed an incident report, he clearly stated that the floor was dry at the time of Ellen's slip-and-fall event. Obviously, I was not present at the time Ellen slipped and fell and could not speak with certainty as to the level of moisture on the floor but rather relied upon her testimony and what I had seen on the store's surveillance video.

Although the store manager's report claimed that the floor was dry, in his deposition he stated that he did not actually get down on his hands and knees to feel the floor and that he could not say with certainty that the floor was not sweating. I was then asked if I was aware of any other individuals who may have slipped and fallen at the defendant's market as a result of a sweating floor, which I was not. In fact, in the discovery documents provided to me the defendant claimed that they did not have any previous slip-and-fall claims in their entranceway.

As my deposition was ending, the defense attorney asked to clarify my recollection of events from my review of the market's surveillance video. He asked me to describe Ellen's appearance and her actions as she walked into the market, whereby I went on to describe Ellen as a middle-aged, slender white woman wearing a floral skirt and white blouse. He then asked what Ellen's hair color was, which I recalled as being blonde. The attorney then asked me if I was certain of her description, to which I said yes. After a long pause, he asked, "Mr. Kendzior, would it surprise you to know that the plaintiff is a heavy-set young black woman with dark hair who was wearing a T-shirt and blue jeans?" Surprised by the question, I replied yes, and that that was not my understanding of Ellen's appearance. I then looked toward Ellen's attorney, who affirmed my description. The attorney then drew my attention to the store surveillance video, which he had pulled up on his laptop; he played a short clip for me. Sure enough, I got Ellen's physical description all wrong, or so I thought.

Upon closer examination of the time stamp imprinted in the video, it was approximately fifty minutes before what I had reviewed. How could this be? Both attorneys were now looking at each other in amazement and fast-forwarded the video to the time I recalled seeing Ellen slip and fall; sure enough, we saw Ellen's slip-and-fall event, just like I remembered. The defense attorney found an earlier slip and fall, which, like Ellen's, had taken place at the shopping cart storage area. What are the chances that there would be two slip-and-fall events just a few minutes apart, both captured on store surveillance video? What I thought was going to be a bad day for me turned out to be a bad day for the defendant's attorney. My deposition quickly came to a close.

In the first instance a store employee is seen attending to the victim, who is shaken but apparently not hurt. The employee and woman are approached by the store manager, who writes down something in a notepad he takes from his front shirt pocket. All three people are then seen leaving the area. There was now no question that the defendant was aware that the floor in their entranceway was hazardous prior to Ellen's visit. Such knowledge is what is called actual notice and is a cornerstone requirement that plaintiffs have the burden to prove in a slip-and-fall case.

HOW DID IT END?

After my deposition was taken, the store manager was redeposed and asked if he recalled the first slip-and-fall event, which he did. He recalled speaking to the young black woman seen in the store's surveillance video, and when asked why he claimed that there were no previously slip-and-fall claims, he said that the woman did not want to file a claim and that he simply recorded the time of her fall and that the floor was wet. In his view if an incident report is not recorded, then the event didn't happen. This is a common practice and legal strategy used to make it more difficult for plaintiffs to prove notice and is akin to the saying "If a tree falls in the forest and no one is there to hear it, does it make a sound?"

Because of the manager's new testimony and that of a known, previously recorded slip-and-fall event captured on surveillance video, the defendant quickly settled the case.

9

SLIPPERY SLIPPERS!

On December 10, 2003, I had received a box containing three pairs of designer slippers from an attorney in Philadelphia, Pennsylvania. In the box was a pair of new, blue slippers, which were attached to a promotional hang tag, and a second, pink, slightly worn slipper. Both the new and worn slippers were identical in style and design and differed only in color and wear. The slippers' outsoles were made of a synthetic polyester or acrylic woven material that was formed into a waffle pattern. The attorney asked for me to render an opinion as to the slippers' level of slip resistance.

It appears that shortly after purchasing the slippers, the plaintiff, Jennifer, experienced a serious slip-and-fall event as she entered her home's ceramic-tiled kitchen. Although the floor was clean and dry, her new pink slippers caused her to slip and fall, and she subsequently landed on her tailbone. Jennifer stated that after landing, she saw stars and could not get herself up and off the floor.

Shortly after receiving the slippers I had forwarded them to Artech Testing, L.L.C., for slip-resistance testing. At the time Artech was a leading independent footwear-testing laboratory used by most of the U.S.-based footwear manufacturers, and it specialized in slip-resistance testing. On February 4, 2004, I had received the first of two reports from Artech Testing. The report detailed the slip-resistance characteristics of each of the slippers as measured by the ASTM F489-96 James Machine test method, which at the time was the most widely accepted test method for determining footwear outsole slip resistance. In fact, the term "slip resistant" is defined as possessing a static coefficient of friction (SCOF) of 0.5 or greater as measured by the James Machine under this test method. Upon testing it was found that each of the slippers had SCOF values far below the industry-accepted standard of 0.5. The pink (worn) slipper had a SCOF value of 0.16 and the blue slipper (new) tested at a value of 0.13.

To confirm the validity of this data, I had requested that Artech Testing conduct a second round of slip-resistance testing, this time using a different test method that is conducted on two types of walkway surfaces, including vinyl composition tile (VCT) and quarry tile.

On February 13, 2004, I had received a second report from Artech Testing, which revealed results similar to those of the previous report. The pink (worn) slipper was tested to have a SCOF value of 0.19 when tested against a dry VCT and 0.43 when tested against a dry American Olean red quarry tile material, which has a rougher surface texture. The blue (new) slippers tested similarly and had values of 0.18 for VCT and 0.40 against the quarry tile. Again, the data revealed slip-resistance values below the industry standard for safety.

Based on this information, it is my opinion that the slippers in question were clearly not slip resistant and posed a serious slip hazard. Furthermore, I believe that the manufacturer's sales literature overstated the safe use of their product, which directly contributed to Jennifer's slip-and-fall event. The literature described "50 ways to wear your slippers" which included such activities as "chasing the kids, making dinner, cleaning the house, washing dishes, and fixing breakfast," to mention a few. To suggest that these slippers could be safely worn to "chase the kids" is a gross overstatement of their product's safe use. In fact, all the recommended household uses as stated in the manufacturer's sales literature would be questionable as it relates to safe use of their product. It is my understanding that Jennifer's slip-and-fall accident occurred while simply walking across her kitchen floor, which was one of the recommended uses as described by the manufacturer's own literature. Had the manufacturer attached a warning label describing the possibility of slipping when wearing their product, Jennifer could have made a more-informed decision as to the safety of the product and its appropriate use.

It was therefore my conclusion that the manufacturer, licensee, and seller had produced and marketed a product that was unsafe for its intended purpose. The slipper outsole offered substandard slip resistance, which in my opinion directly contributed to Jennifer's slip-and-fall event and subsequent injuries. The synthetic waffle-patterned sole is an inappropriate material for a house slipper, and based on independent laboratory test data, such a material should not have been selected as an outsole material.

HOW DID IT END?

The case settled for a six-figure sum.

10

BROKEN TELEVISION!

On September 30, 1999, Fred was returning a television to a local San Antonio, Texas, TV rental store when he suddenly tripped and fell over an elevated display/platform, which was located adjacent to the front entryway door of the store. There was nothing on the display at the time of his fall, and the display platform was covered in low-pile carpeting identical in color to that used on the floor approximately eighteen inches below. There were no guardrails or handrails installed at the front entryway to serve as a visual cue to Fred. Fred's fall resulted in a broken shoulder and wrist . . . and a broken television!

Based on the photographs that were provided to me by Fred's lawyer, it was hard to see the display. Needless to say, as it existed, the carpeted display presented a serious potential trip hazard. It is the industry standard for elevated platforms to have a contrasting color to that of the adjacent floor to attract pedestrians' attention to the object. This is particularly true for retail stores, where incoming customers' attention is being directed toward store displays or merchandise. Magnifying the risk even more was the fact that Fred's vision was impaired, as he was carrying a large television in front of him.

Secondly, given the fact that the store's management chose to locate the elevated display platform directly in front of the front doors, they should have installed a guardrail or handrail to protect an incoming visitor from accidentally tripping over the elevated platform. Installation of pedestrian guardrails is a reasonable safeguard in applications where the entry doors open out onto the exterior walkway and therefore can not assist in directing incoming pedestrian traffic away from interior obstacles.

THE LAWSUIT

While Fred's lawyer argued that the display presented a trip hazard, the store's attorney stated that if it had, how is it that no one else had tripped over it? Fred simply was not watching where he was walking, a fact he acknowledged. The case soon became a question of what responsibility a retailer has to safeguard their guests from a possible trip hazard.

It was my opinion that Fred's trip-and-fall event was both predictable and preventable had the store's management simply located the elevated platform display away from the front entryway or installed protective entryway guardrails. Their failure to do so was therefore a direct cause of Fred's injuries. It is my opinion that the store's management had failed in their duty to exercise good judgment, and in turn, exposed their invited guests to the unnecessary risk of a trip and fall. The store also agreed that it was customary for customers to return merchandise to the store via the front door.

HOW DID IT END?

The case settled for $50,000.

11

MINING FOR PEANUTS!

On October 24, 2002, Joe and a group of friends were out on the town having a good time at the Buckaroo Mining Company bar and nightclub located in Houston, Texas. It was about 1:00 a.m. when Joe and his friends entered the bar, where he soon found himself on the floor near the bar. The bar was packed and the lighting dim. As Joe approached the bar, he felt the crackle of peanuts on the stained wood-plank flooring beneath his cowboy boots. His next step would change his life. Joe's friends told him that he slipped on what they called a "carpet of peanuts." So heavy were the peanuts on the floor, they could not see the wood-plank flooring. Joe's injury resulted in a broken right femur.

Although wood plank is an appropriate walkway material for a commercial establishment, it does require proper maintenance. Walkways, like that of a nightclub bar, take a great deal of abuse from foot traffic, spilled food and drinks, and so on, which, over time, contaminate the walkway surface, causing it to become slippery. This phenomenon is accelerated with the addition of peanuts/peanut shells disposed of onto the walkway.

Having done a little research on peanuts, I learned that approximately 37 percent of the mass of a peanut is comprised of peanut oil; therefore, for every pound (16 ounces) of peanuts being dropped on the floor, 4 ounces of peanut oil is available for contamination (www.woodrow.org/teachers/ciu/1988/peanutlab.html). Furthermore, vegetable oils like peanut oil are difficult to remove from walkway surfaces and require a specialty cleaning solution and procedure.

It was my opinion that due to the continued application of peanuts onto the floor in question, the surface was contaminated and therefore unsafe for pedestrian use. Industries like bars, restaurants, and nightclubs are prone to slip-and-fall accidents, so they should take prevention of such accidents very seriously. Not only did the management of the Buckaroo Mining Company not take this subject seriously, they compounded it by encouraging their patrons to dispose of hazardous substances (i.e., peanuts) onto the walkway. Such practice is not the industry standard, and it is my opinion that the Buckaroo Mining Company did not provide for a reasonable standard of care for their walkways.

The American Society of Testing and Materials (ASTM) has defined the standard of care for pedestrian walkways and has published such a standard, entitled ASTM F-1637-95, "Standard Practice for Safe Walking Surfaces." The scope of this standard states that "This practice covers the design and construction guidelines and minimum maintenance criteria for new and existing buildings and structures" and calls for the following safety guidelines:

Section 4.1.3: "Walkway surfaces shall be slip resistant under expected environmental conditions and use." Section 4.1.4: "Interior walkways that are not slip resistant when wet shall be maintained dry during periods of pedestrian use." Section 4.4.3: "Mats and runners should be provided at other wet or contaminated locations."

The walkway in question violated the guidelines listed above, which is the basis of my opinion that the Buckaroo Mining Company was not in compliance with the standard of care of pedestrian walkways.

Buckaroo Mining Company hired an expert of their own, who prepared a report that stated that the floor in question was reasonably safe and in compliance with the ASTM C-1028 standard for measuring static coefficient of friction (SCOF). I am familiar with this standard and serve on the ASTM committee that authored it. The C-1028-96 method is strictly used for ceramic tile and other similar materials (e.g., porcelain, mosaic tile, etc.) and therefore, the defense expert's opinions were based on the improper reference and use of a test standard. I generally do not like to personally attack the other side's expert, and when asked to rebut opposing counsel's expert's testimony, I limit my criticism to the facts and not personalities.

Furthermore, the defendant's expert claimed that because wood flooring is often treated with oil, such walkways are safe when oily, thus suggesting that wood floors can be made safer by the addition of peanuts, peanut oil, and peanut shells.

THE TRIAL

The defense expert produced a report that contained a page of slip-resistance data he acquired by testing the floor to the wrong standard. His dry surface test data claimed that the wood floor without the addition of peanuts and peanut shells possessed a slip-resistance value of 0.56. However, when peanuts, peanut oil, and peanut shells were added, the slip resistance rose to a value of 0.58. This relationship was also seen when examining his wet slip resistance data. The expert's data further alleged that the uncontaminated wet floor's slip resistance rose from a value of 0.52 to a value of 0.55 when peanuts, peanut oil, and peanut shells were added.

It is my view that the defense expert either did not understand what his data was saying or that he simply failed to closely examine his test results. Had he done so, he would have realized that such conclusions are scientifically incorrect and defy common sense, for which he provides no explanation. Simply put, the slip resistance (i.e., safety) of a walkway is not improved by the addition of peanuts, peanut oil, and peanut shells, but in fact is greatly reduced.

The defendant's expert then referenced a value of 0.5 as being "slip resistant" and had implied that surfaces that possessed such a value would be considered safe. The ASTM C-1028-96 test method does not reference any numerical value nor does it define the term "slip resistant" by way of a COF level. The only ASTM that does such is the ASTM D-2047 standard, which is used exclusively by the floor polish industry and is not applicable to wooden walkways. It was my view that he had confused the two standards and was therefore incorrect in his statement as to the significance of the 0.5 value.

Lastly and perhaps most importantly, the defendant's expert had conducted his evaluation of the walkway on July 14, 2004, nearly two years after the slip-and-fall incident that occurred on October 24, 2002. Such a delay would make any comparisons as to the safety of the walkway in question nearly impossible. It is highly unlikely that the walkway he had tested on July 14, 2004, was identical to the floor the day of the incident, and therefore his conclusions about the safety of the walkway should be questionable as to their accuracy and relevance.

It was my opinion that patrons should not be encouraged to dispose of peanuts and peanut shells onto the walkway. Such practice directly increases the risk of injury due to slipping on a peanut, peanut oil, or peanut shell. Such practice was encouraged by the management of the Buckaroo Mining Company, for which they were liable.

It was my opinion that based on the facts as presented to me at the time, Joe's slip-and-fall event was directly caused by peanuts on the floor, which were known to be present by Buckaroo Mining Company's management prior to the event, and which they encouraged their patrons to throw onto the floor. Joe's slip and fall could have been prevented had the walkway been inspected for hazards and properly cleaned so as to not have a thick layer of peanuts. Unfortunately, the management failed in their duty to exercise good judgment and in turn, exposed their employees and invited guests to unnecessary risk. Had the restaurant's management addressed this hazard prior to the time of Joe's visit, he would not have slipped and fallen.

HOW DID IT END?

The jury found in favor of the Buckaroo Mining Company. Interestingly I have been retained in several similar peanut cases where defendants, usually a country-themed restaurant, provide peanuts in the shell for their customers to eat and discard on the floor only to create a slip hazard.

12

WEDDING CRASHER!

On November 3, 2002, Mary was a guest at her nephew's wedding reception, which was being held at an upscale downtown Dallas banquet room. The banquet room chandeliers were dimly lit, and dinner was about to be served on fine china dinnerware. Mary was a well-known civic leader and dignitary who served on numerous non-for-profit boards and was a founder of one of the most popular Christian youth camps in the Southwest. As Mary walked across the ballroom, she tripped on a curled edge of one of the rugs, causing her to fall forward and break her right arm. At the time of her fall Mary was wearing a pair of dress sandals. Mary was also a highly skilled competitive tennis player.

Located on the ballroom floor were a series of three large woven oriental area rugs. The rugs were of the finest quality and showed their seventy-five-plus years of age. If only the rugs could speak. They would tell you of the city mayors, state governors, senators, and even a few presidents who walked across their surface. They were indeed beautiful rugs, but in need of retirement. The rugs were not only heavily worn but had curled corners and warped edges that made them appear to be more trapezoidal than rectangular.

THE LAWSUIT

I was retained by the plaintiff to render an opinion as to the safe condition of the area rugs. During my inspection of the rugs I was informed
by the ballroom manager that the carpets were installed as they normally were, which was the same as the day of Mary's trip and fall. I had
measured gaps between the carpets from one-half to one-and-a-half
inches in width, which exposed the wooden floor beneath. The rugs
were not adhered to the ground nor were they taped to each other. The
defendant argued that the rugs were in good condition and because of
their age, were expected to display signs of wear, and stated that "that's
what gave them character." The case came down to whether an antique
area rug with character as used on a ballroom floor created an unreasonably dangerous condition.

It is my opinion that multiple failures had occurred that could have
prevented Mary's trip-and-fall event and subsequent injuries. First, it is
my opinion that the management of the ballroom exercised poor judgment in their decision to install the area rugs in such a way that the
edges were not adhered to the floor. Such improper installation served
to camouflage the pending trip hazard of the exposed carpet edge,
especially in the traffic aisle where guests were collected. Evidence of
this comes from the deposition testimony from the ballroom's employees, Alfreda and Martha, who were service attendants and eyewitness to
the event.

Prior to Mary's trip and fall, Martha recalled seeing people catch
their chair legs on the edge of the carpet "two or three times." She also
described seeing the carpet "lift a little when people come in and out
from sitting" and had once told a woman to be careful as she lifted the
edge of the carpet as she pulled out her chair.

Alfreda described the condition of the carpets in her deposition as
being "real stained" and often misaligned. In an effort to straighten out
the carpets, she further described workers having to "walk on them
(rugs) to try to straighten them out." Alfreda also stated that she personally witnessed seeing the rugs develop a "bump or elevation" between
them, and as a corrective measure she had personally "put tape on
them" a couple of times. All the above had taken place prior to Mary's
trip and fall.

Section 4.4.5 of the American Society of Testing and Materials (ASTM) F-1637-95 Standard Practice for Safe Walking Surfaces (item 1) states "Mats, runners and area rugs shall be provided with safe transition from adjacent surfaces and shall be fixed in place or provided with slip resistant backing." The carpets in question did not provide a safe transition from that of the adjacent carpet, were not fixed in place, and therefore were not in compliance with the ASTM guideline stated above.

The use of area carpeting for use in a banquet hall is common; however, when area carpeting is used, it is the standard practice to tape the conjoining edges together to prevent a trip hazard. During my site inspection, I have observed several small pieces of brown duct tape adhered to the backs of the area carpets. This evidence suggests two things; first, that the Tower Club's management acknowledged in the past the need to adhere the carpet seams together, and in fact tried to do it by taping the seams together with duct tape. Unfortunately, on the evening of November 3, 2002, the carpets were not secured to the floor in any way, the result of which was a direct cause of Mary's trip and fall. This fact is substantiated by several of the above-named witnesses to the accident as well as Mary herself.

Furthermore, the likelihood of this impending trip hazard was further compounded by the following contributing factors:

1. The room was dimly lit and made it more difficult to see any trip hazards (e.g., exposed edges).
2. The tablecloths were draped in such a way as to cover the rugs, thus concealing their raised edges.
3. The close configuration of the banquet room tables and chairs created paths/"traffic aisles" for pedestrian access. One such path funneled traffic directly over the exposed carpet edge, thus increasing the likelihood of a trip and fall. It is at this location that Mary tripped and fell.

Based on this information, the management and staff of the banquet hall had in fact created multiple trip hazards that directly caused Mary to trip and fall. Furthermore, the proper taping techniques were known to the banquet hall's management but not used on the night of November 3, 2002. Had the defendant addressed the walkway hazards prior to Mary's visit, it is unlikely that she would have tripped and fallen.

HOW DID IT END?

The case was scheduled for trial and was settled "on the courthouse steps" for an undisclosed sum.

13

BIG-TEX'S SOCCER BARN!

On the evening of June 1, 2012, Nicki was playing soccer with her indoor soccer team, the Orange Crush, when she tripped and fell on an open edge of soccer turf, causing her to tear the ligaments in her ankle.

Nicki was a nineteen-year-old recreational soccer player with a dream of someday being a professional player. She had been to the Big-Tex Soccer Barn previously both as a spectator and as a player. On the date in question Nicki was a player on the Orange Crush soccer team, and as she was exiting the artificial-turf playing field, she caught her right foot/shoe in an elevated section (gap) of artificial turf, which caused her to fall to the ground, seriously injuring her right knee, ankle, and foot. Nicki was wearing a pair of soccer shoes, which were appropriate for use on artificial sport turf.

INDUSTRY STANDARDS

Artificial sport turf like that in question is a carpet-like product, which although used for athletic sporting events, must comply with the industry standards for carpeting as well as sporting fields. The leading organization that governs the standards for both is the American Society of Testing and Materials (ASTM). Technical literature as published by the ASTM STP-1073 entitled "Natural and Artificial Playing Fields Characteristics and Safety Features" requires that artificial sport turf be properly installed, inspected, and maintained to ensure the safety of those who use it. Aspects that directly relate to player safety include but are not limited to turf type, installation technique, cushioning, durability, and maintenance.

A second industry standard that is also published by the American Society of Testing and Materials is entitled "ASTM F-1637-10 Standard Practice for Safe Walking Surfaces." Section 5.3.1 of the standard requires that "Carpet shall be maintained so as not to create pedestrian hazard. Carpet shall be firmly secured and seams tightly maintained. Carpet shall not have loose or frayed edges, unsecured seams, worn areas, holes, wrinkles or other hazards that may cause trip occurrence." Section 5.3.2 of the standard further requires that "Carpet on floor surfaces shall be routinely inspected. . . ."

OPINIONS

Based on the information presented to me at this time, it appears that the seamed section of turf was either improperly installed or damaged during regular play and had separated so as to create a trip hazard. It is my opinion that multiple failures had occurred that could have prevented Nicki's trip-and-fall event and subsequent injuries. First, it is my opinion that the defendant exercised poor judgment in not properly inspecting their playing fields prior to each match. Had they done so, they would have noticed the separated seam and made the necessary repairs prior to allowing players onto the field. It is also my opinion that Nicki acted reasonably as a player, and is not responsible nor expected to identify turf-related defects including bulges, ripples, or separated seams but rather has to rely upon the provider of the facility to ensure such via timely inspections and necessary repairs.

Big-Tex's Soccer Barn's website states "Play Where the Pros Play" and also offers guests to "have your next birthday party at Inwood soccer center," thus suggesting that they cater to a wide range of players, any of which could have experienced the same unsafe playing field as that of Nicki. Given such, the defendant had an obligation to provide a safe playing field to all their invited guests, including Nicki. Had the sport turf/carpet in question been properly affixed, maintained, and inspected prior to each match, it is unlikely that Nicki would have tripped and fallen and injured herself.

It is unclear whether or not the defendant had performed frequent or scheduled inspections of the sporting facility's turf/carpet or whether any of their management or staff was properly trained to identify potential turf-/player-related hazards. However, it is my understanding based on Nicki's deposition testimony that she had previously found artificial turf "bulges" on the playing field in question, whereby she promptly reported such to game officials as well as the defendant's management. Given such, the defendant was aware of previous and ongoing turf-/carpet-related defects/hazards and apparently failed to make the appropriate repairs. Given such, the defendant's failure to provide a safe playing field was in my opinion the direct cause of Nicki's trip-and-fall event.

Based on the evidence presented to me at this time, it was my opinion that the defendant failed to comply with the above-named industry standards, which establish the standard of care for artificial turf–covered recreational playing fields and in doing so created an unreasonably dangerous condition for those individual players of the Big-Tex Soccer Barn, specifically Nicki.

Also, had the defendant's management addressed the above-mentioned turf installation hazard prior to Nicki's visit, it is unlikely that she would have engaged her foot in the open seam to cause her to trip and fall. It is therefore my conclusion that the defendant was negligent in their duty to protect their invited guests from unnecessary harm and was directly responsible for Nicki's fall and related injuries.

HOW DID IT END?

Big-Tex's Soccer Barn went out of business midway through the lawsuit, but Nicki settled with their insurance company for $165,000. Her foot healed but was never the same, and she can no longer play soccer.

14

HE GOT THE POINT!

Back in 2001 I was retained in a case in Chicago where a construction worker named José was working on a job site, and as he was walking around a pile of construction rubble, he tripped and fell. The job site was a renovation of a high-end retail store in the heart of the city. José was working on the third floor, which was only lighted by a string of overhead lightbulbs. The rubble was the result of the demolition crew's previous day's work and consisted of broken bricks, cinder blocks, electrical wire, and so on. José's job was that of a helper, and he was on his way to the dumpster with a wooden door. When José tripped and fell, he landed on the door, and as he raised himself up, he found himself impaled by a piece of metal electrical conduit. The bent and broken piece of conduit went clear through the door and José. In shock, José was approached by his supervisor, Tony, and a few coworkers. Everyone was aghast. Tony knew that if this accident got recorded, he might lose his job. Even worse, José was an undocumented worker, which is strictly forbidden by the City of Chicago trade unions. Tony helped José down to the street level and pointed him in the direction of the hospital some six city blocks away. José, door under his arm and impaled by a piece of metal conduit, walked six city blocks to the hospital. Upon arrival, he was immediately taken into surgery to remove the conduit and to repair his damaged organs. In the recovery room the doctors told José that he was a lucky man. The conduit missed all his critical organs and he should have a speedy recovery. They also told him that had he pulled the conduit out before arriving at the hospital, he would have bled to death.

THE LAWSUIT

José sued the general contractor (GC) who hired him along with four individual subcontractors, including their respective unions. The electricians' union countersued the masonry union, claiming that they were at fault. The masonry union sued the demolition company, claiming that they were responsible, and they in turn sued the GC, who sued all the subcontractors. What a mess! As an undocumented worker, José did not have medical coverage and was responsible for a large medical bill. He had no choice but to seek payment from his employer . . . legal or not!

My testimony addressed the condition of the walkway, which was deplorable. Because José was a worker in the workplace, OSHA rules were applicable. Section 1910.22 of OSHA's Walking and Working Surfaces requires that:

(a) Housekeeping.

(1) All places of employment, passageways, storerooms, and service rooms shall be kept clean and orderly and in a sanitary condition.

(2) The floor of every workroom shall be maintained in a clean and, so far as possible, a dry condition. Where wet processes are used, drainage shall be maintained, and false floors, platforms, mats, or other dry standing places should be provided where practicable.

(3) To facilitate cleaning, every floor, working place, and passageway shall be kept free from protruding nails, splinters, holes, or loose boards.

(b) Aisles and passageways.

(1) Where mechanical handling equipment is used, sufficient safe clearances shall be allowed for aisles, at loading docks, through doorways and wherever turns or passage must be made. Aisles and passageways shall be kept clear and in good

repairs, with no obstruction across or in aisles that could create a hazard.

(2) Permanent aisles and passageways shall be appropriately marked.

Additional safety standards applied, but in the end, it was obvious that the walkway was unsafe. I was deposed by all four of the codefendants, which was quite interesting. What quickly became apparent wasn't whether there was liability, but rather who was going to pay. The union representatives were also present during my deposition, and I later learned they agreed with my opinions. It appeared to me that the union representatives were more upset that the GC hired a nonunion worker than the fact José was hurt.

HOW DID IT END?

José received a settlement from each of the defendants, which was enough to pay his medical bills. Sadly, the law firms representing the various codefendants made more money than José. But that's the system. Tony did indeed get fired but later opened his own company and is now a general contractor.

15

CURBED AT THE ANTIQUE STORE!

I received a call from a defense attorney in South Texas whom I had worked with in the past regarding a woman named Janet, who fell off a curb while leaving an antique store. He later sent me photos of the curb, which showed that it had been painted in red along its face and top edge. I performed a site inspection in November of 2002 and noted that the curb was in compliance with the appropriate walkway safety standards and the local building code.

Janet stated that prior to experiencing her fall she was a bit confused as to where she had parked her car and was preoccupied with trying to spot it as she was exiting the antique store. She admitted that she was not paying attention to where she was walking and simply stepped off the curb. Janet's fall is what is often referred to as an "air-step."

Janet also stated that she was aware of the elevated curb that existed adjacent to the parking lot prior to her fall and that she had walked over it earlier in the day. Janet's fall occurred at approximately 1:00 p.m.; the weather was partially cloudy and it was not raining.

As far as medical conditions go, Janet was in good shape for her age but had poor vision that required her to wear correctional glasses, which she was not wearing at the time of her fall. Although Janet described her fall as a slip, it was most likely a misstep. Based on Janet's description of the accident and the position of her body after she fell, it is my opinion that she fell forward and not backward, thus indicating a trip, stumble, or misstep. Forward falls like that described by Janet are usually the result of a trip or loss of balance rather than a slip. It is

therefore most likely that Janet simply misstepped off the curb and fell into the parking lot, thus injuring herself.

It was my conclusion that the cause of Janet's fall was most likely her preoccupation with locating her car and the impending curb lying directly in front of her. Furthermore, based on my thorough examination, neither the walkway surface, curb, nor apron landing was structurally unsafe and therefore they did not directly contribute to Janet's fall.

HOW DID IT END?

When presented with my report, Janet dropped her lawsuit, saying, "That sounds about right."

16

AN ISLAND OF SAFETY!

It was a drizzly February morning in 2010 when Karen went to her local big box retailer to shop. As she pulled into the store's parking lot and walked into the front entrance, she slipped and fell as she stepped off a carpet mat and onto the highly polished stamped concrete floor. Warning signs were posted; however, the nearest sign was located adjacent to the opposite entrance mat and nowhere near the wet floor hazard where Karen slipped and fell. Karen's injury was very severe. As she slipped, her left leg went out in front of her and her right knee went straight down, fracturing her artificial knee and splitting her femur bone from her knee to her pelvis. The sound of her femur snapping was so loud that the guest greeter thought that someone had shot a gun in the front lobby. Karen was hospitalized for months and was told that her leg might have to be amputated. Store surveillance video revealed the entire episode, including the guest greeter pushing a dust mop in the entranceway minutes before Karen's slip and fall.

INDUSTRY STANDARDS

At the time, there were two nationally adopted consensus standards that define the use, maintenance, and inspection of carpet mats like that used at the front entrance of the big box retail store in question. They include the American Society of Testing and Materials (ASTM) F-1637-02 standard entitled "Standard Practice for Safe Walking Surfaces" and the American National Standards Institute (ANSI) A1264.2-2001 standard entitled "Provision of Slip Resistance for Walking and Working Surfaces." Both the ASTM F-1637-02 and ANSI A-1264.2-2001 standards have been in publication for more than ten years and are the most authoritative safety standards related to walkway safety.

The ASTM F-1637-02 standard, Section 5.1.4, requires that "Interior walkways that are not slip resistant when wet shall be maintained dry during periods of pedestrian use." Section 5.3.2 states that "Carpet on floor surfaces shall be routinely inspected. . . ." Section 5.4 Mats and Runners, Subsection 5.4.1: "Mats, runners, or other means of ensuring that building entrances and interior walkways are kept dry shall be provided, as needed, during inclement weather. Replacement of mats or runners may be necessary when they become saturated." Section 5.4.5: "Mats, runners and area rugs shall be provided with safe transition from adjacent surfaces and shall be fixed in place or provided with slip resistant backing."

Section 6.3 of the American National Standards Institute (ANSI) A1264.1 standard requires that "Mats shall be safely installed so that they do not create tripping hazards." Section 6.4: "Mats and runners shall be routinely inspected and adequately maintained. Damaged mats shall be promptly replaced." Section 6.5 requires that "Procedures shall be established for the placement, maintenance, inspection and storage of mats. Mats or runners shall be stored to prevent curling of edges."

It was my opinion that the mat in question was not in compliance with the two nationally recognized standards stated above.

WINDSOR CORPORATION WHITE PAPER

Industry research has shown that the length of entranceway floor matting is directly proportional to the amount of soil/moisture removed from incoming pedestrians' footwear. According to the report, only 40 percent of the soil brought in via pedestrians' footwear is removed by a 6-foot carpet mat and it takes 36 feet of entrance matting to remove 99 percent of incoming pedestrian foot soil.

OPINIONS

Water migration into entranceway vestibules is common in the retail/ supermarket industry, for which slips and falls resulting from such hazards are a leading cause of both associate and guest injuries. For this reason, retailers including the defendant have adopted written walkway inspection programs or "floor sweep" procedures, which require store management and their associates to conduct frequent inspections of their floors to reduce slip-and-fall injuries.

According to the big box retailer's policy for "Safety Sweeps": "A safety sweep is when Associates sweep or walk the floor and look for and correct potential hazards. Correct hazards as soon as possible. If you need help, guard the floor and ask another Associate to get help to take care of the problem. Use Safety Tents to warn Associates and Customers of slippery floor." Furthermore, according to the big box retailer's "guest greeter" job description, such employees are responsible for "Customer Satisfaction, Safety, and Loss Prevention." It is unclear as to whether the associates of the big box retailer were conducting floor sweeps as outlined in the big box retailer's Safety Sweep policy; however, based on the surveillance video, it appears that they were not.

The location of Karen's slip and fall was a heavily trafficked area of the store, in plain view of the defendant's employees who were responsible for inspecting the floors and entrance mats. Moments before Karen entered the big box retail store's vestibule and slipped and fell, a big box retailer associate, presumably the guest greeter, was seen on surveillance video mopping the entrance floor with a dust mop. Dust mops such as that used by the big box retailer employee are usually treated with a chemical that accelerates their ability to attract and hold dry soils (e.g., dust) and repel moisture. Given such, dust mops are not intended to be used for removing water. The process of removing water should be done using a dry mop and bucket, paper towels, or towel.

The stamped concrete in the big box store's entrance was coated with a high-gloss acrylic coating that had a very low coefficient of friction (COF). Although stamped concrete like that used in the big box retail store's vestibule is common in the retail industry, it should not be coated via low-traction acrylic finish. Stamped and finished concrete does not provide an adequate level of slip resistance when wet and should not be used in retail vestibules where the migration of water, snow, and so on is to be expected.

It was my opinion that the length of entrance matting was inadequate for the size of the entranceway and should have been replaced with either walk-off carpet tile or larger/longer carpet mats. Most people see floor mats as islands of safety, where as they step off a mat, the degree of traction will decrease. However, this was different. Each time an incoming guest entered the store, small quantities of water were being transferred from the saturated mat to the adjacent floor, thus creating a serious wet floor slip hazard. Although the guest greeter was seen pushing a dry mop on the floor, all she was doing was making the hazard larger. At no time did any big box retail employee inspect either of the two entrance mats. Based on the photographs taken shortly after Karen's slip and fall, the floor mat in question became wet or saturated from migrating rainwater, which was then transferred onto the adjacent concrete walkway, thus creating an unreasonably dangerous condition. It is my opinion that Karen's slip and fall was the result of her stepping off the saturated floor mat onto a wet floor.

It is my opinion that the big box retailer failed in their duty to provide a safe walking surface to their invited guests, including Karen, and that the big box retailer's employees failed to properly remove such hazardous condition or properly warn of its existence. Furthermore, it is my opinion that Karen's slip-and-fall incident could have been prevented had the big box retailer's management and their associates, specifically the guest greeter seen in the surveillance video, followed the procedures for walkway safety as prescribed by the big box retailer's safety manual, which includes proper posting of warning signs and proper hazard removal. Had the big box retailer's management properly addressed the known slip-and-fall hazard prior to the time of Karen's visit, she would not have slipped and fallen.

HOW DID IT END?

The big box retailer's attorney did a pretty poor job when he took my deposition, and he paid the price at trial. Prior to trial, the big box retailer offered a settlement of $5,000. The jury awarded Karen more than $600,000.

I have had many cases representing plaintiffs against this same retailer, and I have seen a pattern of disregard for customer safety. Although they have good written policies, their employees don't follow them and in turn expose their invited guests to what are otherwise easily preventable slip hazards.

17

BLEACHER FALL!

In 2004 I was retained in a matter involving a young boy by the name of Evan, who on June 17, 2003, fell from the bleachers at the city-owned football stadium located in a south Texas city.

Evan's fall occurred on the side stairway near the rear (top) elevated walkway area of the bleachers. The bleachers were constructed of steel and had staggered wood and aluminum steps. Guardrails were installed along the perimeter of the bleachers; however, only the rear, elevated portion of the guardrail was protected with metal chain-link fencing. Evan's mom recalled seeing Evan slip on the metal seating portion of the bleachers, whereby he fell through the bottom horizontal guardrail, landing on his head approximately thirty feet below and fracturing his skull. Since his fall, Evan has been in a coma. Warning signs were not posted on or adjacent to the bleachers. No form of slip-resistant material (e.g., grating) was present on the walkways or steps.

Although both wood and aluminum planking are appropriate walkway materials for recreational bleachers, they do not possess identical slip-resistance characteristics. While the aluminum steps depicted in the photographs are clean and clear of defects and relatively new, the wooden steps and risers demonstrate varying degrees of age and wear. Such wear directly affects the material's slip resistance and is often the cause of slips, trips, and falls.

Studies show that the static coefficient of friction (SCOF) values of wooden plank walkways are often higher than those of aluminum walkways, thus causing a significant transition in slip resistance when installed side-by-side as in the above-named case. Such transitions are to be avoided and designed out of any public walkways. The American Society of Safety and Materials (ASTM) guideline F-1637-95 entitled "Standard Practice for Safe Walking Surfaces" states such. Section 4.7.1 of this standard states that "Exterior walkways shall be maintained so as to provide safe walking conditions." Section 4.7.1.2 states that "Exterior walkway conditions that may be considered substandard and in need of repair include conditions in which the pavement is broken, depressed, raised, undermined, slippery, uneven, or cracked to the extent that pieces may be readily removed."

This standard also addresses the need for stairs to be constructed and maintained safely. Section 6.1.2 of the standard states that "Steps nosings shall be readily discernible, slip resistant, and adequately de-marcated. . . ."

Finally, Section 10 of this standard, entitled "Warnings," states that "The use of visual cues such as warnings, accent lighting, handrails, contrast painting, and other cues to improve the safety of the walkway transitions are recognized as effective controls in some applications. However, such cues or warnings do not necessarily negate the need for safe design construction."

It was my opinion that the stadium bleachers were not constructed in compliance with the above-mentioned standard and presented a rea-sonable risk of harm to anyone using them. Furthermore, the walkway surface was not uniformly slip resistant and thus created an undue risk of slipping. Although the aluminum steps had a ribbed surface pattern, the wooden steps did not. The wooden steps did not have any slip-resistant safety strip applied to their surface to improve their safety. Finally, none of the stair nosings were properly marked with a contrast-ing color as required under this standard as well as state and local building codes. It is therefore my opinion that the exterior stairway was improperly constructed and in need of repair.

The industry standard for erected steel bleachers is such that all exterior walkways should be protected by a forty-two-inch screened handrail. Although the bleachers did have a forty-two-inch screened guardrail at the rear most elevated portion of the grandstand, they did not provide such a protective guardrail along the perimeter of each walkway.

Such failure is a violation of both the state and local building code, which was at the date of the accident under the jurisdiction of the International Code Councils (ICC) International Building Code (IBC). Protective guardrails are required for all walkways that are elevated more than thirty inches in height. Such guardrails must also be constructed to prevent accidental injury due to a fall through the guardrail. Section 1003.2.12.2 of the IBC entitled "Opening Limitations" states that "Open guards shall have balusters or ornamental patterns such that a 4-inch-diameter (102 mm) sphere cannot pass through any opening up to a height of 34 inches (864 mm). From a height of 34 inches (864 mm) to 42 inches (1067 mm) above the adjacent walkway surfaces, a sphere 8 inches (203 mm) in diameter shall not pass."

The side guardrails named above were not in compliance with the building code requirement for safety and therefore presented an unreasonable risk of harm to individuals, especially small children like young Evan, who could foreseeably be able to pass through the large guardrail opening.

It was therefore my opinion that the city failed to exercise ordinary care in maintaining the guardrails and walkways of the stadium and failed in their duty to protect the public from unnecessary risk of injury. Had the city simply installed the same metal chain-link material along the side guardrails as they did along the back guardrail, Evan would not have fallen through the guardrail, in turn injuring himself.

HOW DID IT END?

Evan's family received a seven-figure settlement from the city. Evan has recovered but has permanent physical disabilities.

18

INJURED AT THE VET'S OFFICE!

In May of 2009 I was retained by a woman named Lisa, who upon walking to her pet's animal hospital, tripped and fell near the main entrance, injuring her shoulder and head. The walkway surface was a pathway that consisted of eight 18-by-18-inch precast concrete stepping stones surrounded by dirt and pea gravel. The space between the stepping stones was not uniform and averaged in space between eight and twelve inches apart. The pathway was wet from a recent light rain. No warning signs or handrails were present at the pathway.

It was my opinion that the owner of the animal hospital exercised poor judgment in not providing an appropriate entrance walkway to their facility. The use of precast concrete pavestones surrounded by dirt and gravel is not an acceptable walkway surface for a commercial facility.

Based on the photographs provided to me, it appeared that the concrete pavers were unlevel, ranging in height from approximately one-half inch to one inch above the surrounding dirt and pea gravel. Such changes in walkway level are considered walkway hazards and often lead to trip-and-fall events. Elevations in texture, like that I had observed at the location where Lisa had fallen, are a direct violation of the American Society of Testing and Materials (ASTM) F-1637-02 "Standard Practice for Safe Walking Surfaces." Specific violations of the standard include:

Section 5.1 General, Subsection 5.5.1, requires that "Walkways shall be stable, planar, flush, and even to the extent possible. Where walkways cannot be made flush and even, they shall conform to the requirements of 5.2 and 5.3."

Section 5.1.2: "Walkway surfaces for pedestrians shall be capable of safely sustaining intended loads." Section 5.1.3: "Walkway surfaces shall be slip resistant under expected environmental conditions and use. Painted walkways shall contain an abrasive additive, cross cut grooving, texturing or other appropriate means to render the surface slip resistant where wet conditions may be reasonably foreseeable." Section 5.2 Walkway Changes in Level, Subsection 5.2.1, requires that "Adjoining walkway surfaces shall be made flush and fair, whenever possible and for new construction and existing facilities to the extent practicable." 5.2.2: "Changes in levels of less than 1/4 in. (6 mm) in height may be without edge treatment shall be beveled with a slope no greater than 1:2 (rise/run)."

Section 5.2.4: "Changes in levels greater than 1/2 in. (12 mm) shall be transitioned by means of a ramp or stairway that complies with applicable building codes, regulations, standards, or ordinances, or all of these." Section 5.7 Exterior Walkways, Subsection 5.7.1 requires that "Exterior walkways shall be maintained so as to provide safe walking conditions." Section 5.7.1.1: "Exterior walkways shall be slip resistant." Section 5.7.1.2: "Exterior walkway conditions that may be considered substandard and in need of repair include conditions in which the pavement is broken, depressed, raised, undermined, slippery, uneven, or cracked to the extent that pieces may be readily removed." Section 5.7.2: "Exterior walkways shall be repaired or replaced where there is an abrupt variation in elevation between surfaces. Vertical displacements in exterior walkways shall be transitioned in accordance with 5.2."

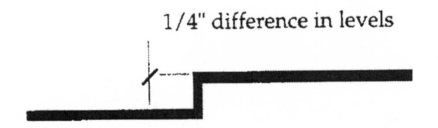

1/4" difference in levels

Further violations include those published by the American National Standards Institute (ANSI) A 117.1.1998. Section 303.2 of this standard requires that "Changes in level of 1/4 inch (6 mm) high maximum shall be permitted to be vertical." Section 303.3 requires that "Changes in level between 1/4 inch (6 mm) high and 1/2 inch (13 mm) high maximum shall be beveled with a slope not steeper than 1:2." Section 303.4 requires that "Changes in level greater than 1/2 inch (13 mm) shall be ramped and shall comply with Section 405 or 406" (ramp section).

Finally, the animal hospital failed to adequately warn Lisa of the walkway hazard. Section 11.1 of the ASTM F-1637-02 standard requires that "The use of visual cues such as warnings, accent lighting, handrails, contrast painting, and other cues to improve the safety of

303.3 Beveled. Changes in level between $^1/_4$ inch (6 mm) high minimum and $^1/_2$ inch (13 mm) high maximum shall be beveled with a slope not steeper than 1:2.

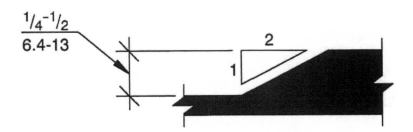

walkway transitions are recognized as effective controls in some applications. However, such cues or warnings do not necessarily negate the need for safe design construction."

It was my opinion that the pathway in question was not in compliance with the national standards listed above. Furthermore, it is my opinion that the pathway in question was an inappropriate pedestrian walkway. Such pathways may be appropriate for residential use or in locations such as parks or nature preserves, where individuals would anticipate an irregular walking surface, but are inappropriate for a commercial business such as an entrance to an animal hospital. Walkways, such as that provided by the defendant, which are designed to connect parking lots to entranceways, should be firm, stable, and slip resistant to provide safe access to those invited guests who would be expected to traverse such walkways. It is common for guests entering or exiting such commercial establishments to be preoccupied, and often they do not pay attention to the surface they walk on. The use of dirt and pavers as an entrance walkway presents an elevated risk of a trip and fall due to the irregular nature of the surface and surrounding material (i.e., dirt and gravel). The pathway in question should not have been provided as a connecting walkway between the parking lot and the building's entrance but rather should have been constructed of concrete like that of the surrounding walkways.

It is my conclusion that the pathway in question posed a significant safety hazard to anyone using it. Based on this information the animal hospital failed in their duty to provide reasonable care for their invited guests to protect them from unnecessary risk of a trip and fall. Had the animal hospital addressed the pedestrian pathway hazard prior to Lisa's visit, it is unlikely that she would have fallen and seriously injured herself.

HOW DID IT END?

The hospital's insurance company settled with Lisa for $300,000.

19

HALFTIME INJURY!

In 2010 I was retained by an attorney who was representing a young man named Matthew who, upon entering the restroom at a major NFL football stadium, slipped and fell on a floor that had flooded from an overflowing commode. According to Matthew, he and his father were season ticket holders and were attending a football game when he left his seat and walked down a concrete ramp that led to the men's restroom. Upon entering the restroom Matthew had to make a sharp left turn around a concrete wall. It was as he was making the turn that his left foot slipped out beneath him, jamming his left foot into the wall and fracturing his ankle. Shortly after Matthew's fall a large group of people began to enter the restroom, including Matthew's father, who helped him out of the restroom. Matthew was transported by ambulance to a local hospital, where he was diagnosed with a fractured ankle and fibula. Prior to his fall Matthew was a competitive surfboarder; his injury would prevent him from playing his sport.

Having reviewed the case file, two things stuck out. The first was, how did the floor become flooded? Secondly, who was responsible for barricading the men's room entrance as to prohibit entry? The stadium hired contracted workers to monitor their facility, including their restrooms. The restroom monitor for the restroom Matthew had entered was not in the restroom at the time but rather was watching the game. His failure to monitor the restroom was a clear dereliction of duty. Shortly after Matthew filed his lawsuit, the restroom monitor died, making it impossible to depose him.

The stadium was in a state where expert reports are not required, so two years after being retained, I was summoned to trial. The day before I was to testify I requested that the defendant permit me to perform a site inspection of the restroom, and they complied. I arrived at the stadium and was greeted by the stadium's risk manager and two security guards who, after signing me in, escorted me on a golf cart to the restroom. My inspection revealed that the restroom floor was a poured resin–based material, which generally provides a high degree of slip resistance; however, the location near the entrance wall showed signs of heavy wear. Because this was a main path of entry, I was not surprised. Furthermore, individuals entering the restroom would have to pivot their foot to the left, which explains the unusually heavy wear. After a brief period, I exited the restroom and began to walk up the ramp toward the stadium seating, whereby the risk manager stopped me. I told him that I wanted to retrace Matthew's path of travel as he entered the door, but the risk manager said that I was only granted permission to inspect the restroom. I then said that as a football fan, I wondered if I could look at the field, and again I was denied.

The next day I was called to testify. Most of the questions asked of me dealt with the standard of care for restrooms, and I went into detail. However, the turning point was completely unexpected. When asked if I could complete my site inspection the previous day, I responded no. I was not permitted by the stadium's risk manager to walk the path that Matthew had walked, and therefore my inspection was incomplete. I then went on to say that I had never been to the stadium, and as a football fan I wanted to look at the field, and I was also denied. When asked if my inability to look at the field impacted my opinions, I said, "No, but that's not the way we treat folks in Texas."

HOW DID IT END?

The jury found in favor of Matthew and awarded him $1.7 million, the largest jury verdict for a slip and fall in the state's history. Much of the award was based on Matthew's lost income as a professional surfboarder. When the jury was polled after the trial, many of them complimented me and said that they were unhappy with the way the risk manager treated me when I asked to see the field. Jurors are people like you and me, and if they are undecided about a verdict, they often base their opinion on who they liked and who they didn't. They liked Matthew but did not like the risk manager, and my short story was the straw that broke the camel's back. The moral of the story is, don't piss off the jury!

20

SIGN OF THINGS TO COME!

In July of 2002 I was retained by an attorney in Oklahoma who was representing a man by the name of Frank, who air-stepped off a curb ramp located in the parking lot of a local restaurant and bar. It appears that Frank had spent a few hours at the bar, and as he left the building to go to his car, he began walking across the sidewalk toward the parking lot where his car was parked. Unbeknownst to him was an unmarked curb ramp that he did not see; he stepped off the sidewalk and down onto the recessed ramp, causing him to lose his balance and stumble forward, whereby he reached out to catch himself on a disabled-parking sign. The sign was made of aluminum and was mounted approximately three feet above the sidewalk level. As Frank reached out with his hand, he caught the edge of the sign between his middle fingers, slicing his hand from his fingers to his palm.

The facts of the case were:

1. Frank's trip and fall occurred on a recently installed pedestrian ramp located in a strip shopping center sidewalk.
2. The accident occurred in the dark and during the early morning hours.
3. The lighting conditions adjacent to the handicap ramp were poor.
4. Neither the ramp nor curb adjacent to the ramp was painted or otherwise colored with a contrasting color from that of the sidewalk.

5. A concrete-filled steel column measuring 33.5 inches tall was located six inches from the side of the ramp.

6. A second concrete-filled steel column of the same height was located approximately fifty-two inches from the side of the ramp, to which a metal reserved-parking sign was mounted.

7. The ramp was constructed as a curb ramp with steep edges rather than a handicap ramp with flared sides. The ramp was not painted, stained, or colored in any way from that of the adjacent pedestrian walkway.

8. Neither the property owner nor the tenant(s) occupying the shopping plaza had acquired the necessary building permits in advance of constructing the ramp or posting the reserved parking signs as required by the city.

Frank sued both the proprietor of the restaurant and the owner of the strip shopping center for failure to provide a reasonably safe sidewalk.

It is my opinion that multiple failures had occurred that could have prevented Frank's air-step and fall. First, not painting the ramp and adjacent curb with a contrasting color from that of the sidewalk made it unrecognizable as a potential tripping hazard. Because both the ramp and the ramp curb were not properly identified with a contrasting color, they were out of compliance with the city's building code as well as federal guidelines and standards established under the Americans with Disabilities Act (ADA), the Standard Building Code (SBC), the International Code Council (ICC), Building Officials Code Administrators International (BOCA), and the American National Standards Institute (ANSI).

Secondly, the placement of the reserved parking sign(s) were too low and in violation of the city's building code and were located in such a way that they posed a safety hazard to persons using the pedestrian walkway. The placement of the reserved parking sign(s) also violated the ADA guideline for "Protruding Objects" as well as the ANSI A-117.1 standard for Accessible and Usable Buildings and Facilities. Furthermore, because the sign was constructed of thin-gauge metal with a sharp edge, it posed a safety hazard to persons using the pedestrian walkway.

Based on this information, it is clear that both the pedestrian ramp and adjacent curb were improperly constructed and were not recognizable as potential trip hazards. Had the property owner and/or tenant(s) contacted the city to attain the proper building permit in advance of construction, they would have been informed of the fact that the ramp they were planning to construct would not be in compliance with local, state, and federal guidelines for a handicap ramp. This is also true for the placement of the reserved parking sign(s). Had the property owner and or tenant(s) attained the necessary sign permit, they would have been informed that such placement was unacceptable and potentially dangerous to persons using the pedestrian walkway. Had the property owner and/or tenant(s) not placed the reserved parking sign(s) low to the ground and directly in the path of persons using the pedestrian walkway, the injuries sustained by Frank would not have been as severe. It is therefore my opinion that the owners of the shopping center along with the tenant(s) responsible for posting the reserved parking signs were negligent in their duty to protect the public, specifically Frank, from unnecessary risk and were directly responsible for his injuries.

THE TRIAL

Prior to trial, the owner of the restaurant and bar settled with Frank for an undisclosed sum, leaving the owner of the strip shopping center as the last standing defendant. This was one of the first trials that I had testified at, and I must admit I was a bit nervous. After an hour or so of trial preparation I found myself in front of twelve good people from a small Oklahoma town. It was clear they did not want to be there and wanted to get back to their lives. My testimony was fairly brief and focused on the standards that define a reasonably safe ramp.

What came up at trial is that Frank had been drinking that night and consumed approximately six drinks. It was also revealed that prior to going to the bar, Frank had been smoking pot. With this testimony I thought there is no way this jury will find in his favor. The law requires business owners to provide a reasonably safe walkway. What I did not expect was what may be considered reasonably safe for a sober person may not be considered reasonable for a person who is inebriated.

HOW DID IT END?

The jury awarded Frank $150,000. They believed that the owner of the strip shopping center knew that one of his tenants was a bar that served alcohol and therefore had knowledge that patrons of the bar may from time to time be inebriated; therefore, he had to provide a reasonably safe environment. In short, even those who may be a bit drunk or stoned should be accommodated, which the defendant failed to do. Interestingly, the defendant was a local attorney.

21

CLOWNING AROUND!

In June of 2003 I was retained by an attorney representing an elderly woman named Diana, who was injured while attending the circus. As Diana was walking down the wooden bleachers, she misstepped off the bottom step. Unlike the other steps, which were unpainted plywood, the last step was painted gray; Diana fell forward, injuring her face and hands.

In July of 2003 I went to the circus to inspect the wooden platform steps. What I found was:

1. The platform was constructed of plywood and measured approximately three feet deep by eight feet in width and six inches in height.
2. The bottom platform step was painted gray and had exhibited signs of edge wear.
3. The circus floor was dark gray–colored concrete.
4. The circus was dimly lit (as one would expect), which made it difficult to see the bottom gray platform step.

It is my opinion that the event coordinator exercised poor judgment in not properly inspecting the temporary walkway fixtures, specifically the elevated wooden platform steps, for safety. Had the steps been properly inspected, the event managers would have seen that the edges of each platform were not properly marked with a contrasting color to that of the adjacent walkway. This fact, combined with the fact that the accident took place under low-level lighting conditions, only served to fur-

ther camouflage the impending trip hazard. The attached photos were taken with a camera flash and without a night-vision camera like that of the circus. Although the platform is clearly visible when the room is well lighted, the bottom step becomes nearly invisible in the dark. The event coordinators were aware that the show was to take place in near-dark conditions and that guests would be migrating up and down the walkways.

The standard practice for walkways containing low-profile steps is to identify the leading edge(s) of each step with a contrasting color. This standard can be found within the following sections of the American Society of Testing and Materials (ASTM) F-1637 Standard Practice for Safe Walking Surfaces. Section 6.1.2: "Step nosings shall be readily discernible, slip resistant, and adequately demarcated. Random, pictorial, floral, or geometric designs are examples of design elements that can camouflage a step nosing." Section 6.2.1: "Short flight stairs (three or fewer risers) shall be avoided where possible." Section 6.2.2: "In situations where a short flight stair or single step transition exists or cannot be avoided, obvious visual cues shall be provided to facilitate improved step identification. Handrails, delineated nosing edges, tactile cues,

warning signs, contrast in surface colors, and adjacent lighting are examples of some appropriate warning cues."

The platform in question was not in compliance with the above-mentioned standards nor was the walkway in compliance with the Americans with Disabilities Act (ADA) or the American National Standards Institute (ANSI) guidelines that serve as the basis of walkway design for most state and local building codes.

It is my opinion that the platform in question was improperly marked and posed a significant safety hazard to anyone using it. Furthermore, the cost to bring the platform into compliance was marginal. For under $10 the event coordinators could have purchased a roll of yellow safety tape to mark the edges of the platform, making it readily identifiable to anyone using it. Because the bottom step was very similar in color to that of the concrete floor beneath it, the event coordinators should not have used a gray-painted step at the bottom landing but rather a step whose color would contrast with the concrete floor.

Based on the facts of the case, the defendant failed in their duty to protect their invited guests from unnecessary risk of a fall and were not in compliance with the standard of care as it relates to the walkway's safety. Had the event coordinators addressed the pedestrian walkway hazards prior to Diana's visit, it is unlikely that she would have tripped and fallen and in turn injured herself.

HOW DID IT END?

Although the defendant's attorneys tried to turn her claim into a three-ring circus, Diana's didn't clown around, whereby Diana settled her case for an undisclosed amount.

22

STAIRWAY TO HEAVEN!

It was May of 2013 when I received a call from Greg, the facilities director of a megachurch located in Houston, Texas, asking if I could come down and test their stairway. Although they had not had a reported slip and fall, they were receiving complaints from their guests that a stairway was very slippery. I told the church that I would be happy to help and that I was planning to be in Houston the following week for another project; I would come by to look at their stairway. Greg said to meet him at his office, which was in the church. Upon arrival, I was amazed by the massive size of the building and upon entry was even more amazed by how clean everything was. This was not just a church but a stadium. Greg gave a tour of the facility, which I thought I would need a map to find my way around. Upon arrival at the stairway in question, Greg told me that several of the guests were elderly and found the stairs to be a bit slippery. I began to test the stairs and not only found them to be safe but in near-pristine condition. The steps were made of terrazzo, which had a high-gloss finish. I asked Greg how he kept the building in such good condition, and he said, "Follow me." We went back to Greg's office, and on the way we passed about a dozen or so maintenance people. Greg told me it was his and the pastor's utmost concern that their guests always have a positive experience, and that begins when you enter the front doors. They had hundreds of first-time visitors and wanted to ensure their complete and total satisfaction, which meant that any complaint, like that about the stairs, should be addressed immediately.

Although the stairs were well within the industry standard for slip resistance, I mentioned to Greg that he may want to consider using a lower-gloss floor finish on them to make them a little easier for their elderly guests to navigate. He agreed, and before I left he asked if I would like to see the auditorium. Upon entering the stadium my attention was immediately directed to the stage, which served as the focal point of the building. Surrounding the front of the stage was a series of two semicircular wooden steps, which were highly polished. I asked Greg if anyone used the steps to get to the stage, and he said "absolutely not"—the steps were not to be used to get onto the stage, and no one could use them. I then asked him how it was that with thousands in attendance, everyone would know not to use the steps. He answered, "They know." Although beautiful, the problem with the steps was that they had a heavy layer of wood polish on them like that of your wood dining-room table, which would make them very dangerous to walk on. Greg agreed and asked me if this was something to be concerned about. I answered yes. Stating that although everyone knew not to use the steps, you never know who may find themselves forgetting. Greg smiled and said that he would take my recommendation to the pastor.

It was about a year later when I was watching a service from the church on TV and smiled. The stairs were gone and were replaced with a lighted-edge short wall.

"The prudent see danger and take refuge, but the simple keep going and pay the penalty." —(Proverbs 27:12)

23

NAILED!

On November 8, 2008, I had received a pair of black, women's size 7.5 designer ladies' boots forwarded to me by an attorney in New York. The boots were in new condition and did not display any signs of wear but had a major safety issue. They were constructed as to have a nail penetrating into the shoe right near the ball of the foot.

What I soon learned was that the plaintiff, Denée, purchased the boots on January 3, 2007, from her local Trident department store located in a suburb of New York City. Denée had purchased the boots without trying them on, and later in the day, as she put on the right boot, she tore a deep gash on the sole of her foot. She then had to remove it, which required her to tear through her skin a second time. Denée's wound soon became infected and left her with permanent nerve damage.

Denée had no idea what had caused such excruciating pain until she removed her foot and noticed a sharp nail protruding into the shoe cavity. Shortly after receiving the boots I had conducted a visual inspection of each boot and found the following:

1. The boots' uppers were made of a synthetic, "leather-like" vinyl compound with a foam lining.
2. The soles were constructed from a synthetic polyvinyl chloride (PVC) compound, which was chemically bonded to the insole and upper.

3. The boots' heels measured 3.25 inches in height and were made of synthetic material designed to appear as wood and were capped with a nylon heel tip.

4. The right boot contained a sharp, nail-like object, which projected into the interior foot bed of the right boot. The nail-like object was centered in the foot cavity and located approximately 2.75 inches from the toe of the boot.

5. The nail-like object penetrated the foot cavity by approximately 1 centimeter (cm).

After performing a visual inspection of the boot(s) I had forwarded the right boot on for X-ray imaging. The X-ray images of the boot confirmed the following:

1. The sharp object was in fact a nail that measured approximately 2 centimeters in length and was bent at a right angle.

2. The nail was sandwiched between the PVC outsole and the insole and protruded into the foot cavity by approximately 1 centimeter.

3. The boot in question contained no other nails.

There was no doubt that it would be impossible for an individual to have placed the nail in question through the insole without puncturing the outsole, which eliminated the possibility that the shoe was tampered with while on the shelf of the Trident store. The fact that the nail was bent at a ninety-degree angle, centered in the foot cavity, and sandwiched between the two layers of the outsole and insole indicates that the nail was placed in the boot during the manufacturing process.

Furthermore, the boot in question was manufactured in such a way as to not require the use of nails, therefore eliminating the possibility that a stray nail accidentally became lodged in the sole. This fact is evidenced by the X-rays, which clearly do not show the presence of any other nails. The outsole and insole of the boot in question were manufactured via a chemical bonding method (i.e., gluing), which did not require the use of mechanical fasteners such as nails. Based on this information, it is my opinion that the nail in question was intentionally placed in the boot during the manufacturing phase. It was most likely a malicious attempt by someone at the point of manufacture to intentionally harm the eventual wearer of the boot.

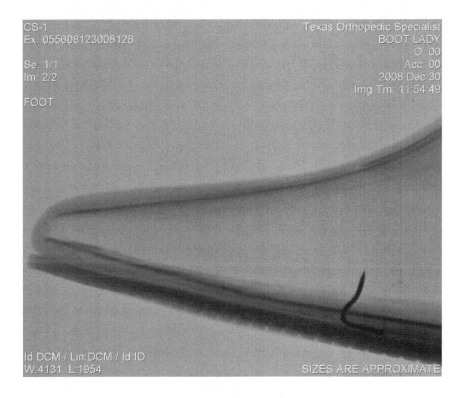

Had the manufacturer and/or their agent properly inspected the defective boot prior to shipment, the nail would have been easily found. It was therefore my conclusion that the manufacturer of the designer ladies' boots, along with the retailer, Trident Stores, along with any third-party agent, was negligent in having offered for sale a product that presented a latent and serious safety defect. The boot in question clearly presented a risk of injury to anyone who would place the boot on. Denée had acted in a reasonable manner and was not responsible for the injuries she incurred as the result of wearing the defective boot.

The Trident Stores' buyer, Carl, testified in his deposition that his company received the order to produce the designer ladies' boot from a well-known fashion designer and in turn forwarded orders to the factory based in Taiwan. The manufacturer was a Trident Stores–approved factory and received orders for the designer boots from Trident Stores buyers. The Taiwan factory was the only factory used to produce the designer ladies' boot and was the source of the defect.

Carl also stated that there was no written agreement between his company and the designer boots' agent, Megashoe, and that his company provided Megashoe with the manufacturing requirements supplied to them by the fashion designer. Carl also stated that he, along with other employees of his company, visited the Taiwan factory to "check the quality level"; however, it was "not his normal regimen" to do so. Carl stated that his company had provided quality assurance documents or standards for Megashoe and that "Trident has its own set of specifics" and would inspect "every order before it left Taiwan." Carl could not recall or produce "Trident's documents" he allegedly passed on to Megashoe and said that he had no documentation, saying "it was a long time ago" and he wasn't sure if they retained the information. Carl stated that he was unfamiliar with the Taiwan factory's testing process but understood that Megashoe "has a QC team in the factory watching production." Furthermore, it was Carl's understanding that Megashoe "looks at each pair of boots," and he felt that his company's quality assurance, inspection, and testing procedures were "sufficient"; however, he was unable to disclose exactly what the quality assurance, inspection, and testing procedures actually were. Finally, Carl stated that a Megashoe company representative told him that it was "impossible" that a tack would be in the designer ladies' Gardenia boot, since that boot did not require the use of tacks (nails).

Rick, the corporate representative from the fashion designer, was under the impression that the Taiwan factory and Megashoe were the same company; he "had never heard of Megashoe," "he had no direct interaction with the factory," and he had never actually visited the factory that produced the designer boots. He stated that the Taiwan factory was responsible for inspections and that his company had not been involved in the testing and inspection of the designer ladies' boots.

Rick stated that the fashion designer had no quality control measures in place for the designer ladies' boot and that he was unaware of any industry or government shoe standards. He further stated that he was unaware of any inspection certificates that were supplied by Megashoe and that to his knowledge there was no written agreement between the fashion designer and Trident Stores.

THE LAWSUIT

Having read all the evidence, it was clear that each of the defendants had "dropped the ball" as it relates to ensuring that the footwear products they produced and sold to members of the public were safe for use. Both Carl and Rick stated that they did not have any written quality control or inspection measures in place as related to the production of the designer ladies' boots, and assumed that the other party was providing oversight for such. It was Carl's understanding that the designer boots' agent, Megashoe, the Taiwan factory, and Trident Stores were providing timely inspections and ensuring that the quality control requirements were being enforced, while Rick assumed that Megashoe was providing such. Neither of the defendants could produce any written documentation as to the specific measures and responsibilities by which designer ladies' boots were to be manufactured nor did they have any written agreement between themselves outlining such responsibilities.

It was my opinion that both codefendants had "passed the buck" to each other and in turn, denied that they were individually responsible for the quality control and inspection measures for the designer ladies' boots, for which each of them was directly responsible. Had the defendants provided a thorough inspection of each pair of designer ladies' boots prior to shipment, the embedded tack (nail) located in Denée's boot would have been caught by the inspector(s) and discarded. Unfortunately, the boot containing the embedded tack (nail) was not inspected and was allowed to progress through the production and distribution chain only to result in direct injury to the consumer.

For decades, manufacturers and producers of footwear have relied upon numerous industry standards governing the manufacturing, inspection, and testing of footwear. These standards have served an important role in assisting footwear manufacturers, producers, and retailers with the necessary guidelines by which to standardize both the quality and safety of the footwear products they produce and market. The major standards-developing organizations producing such standards include the American National Standards Institute (ANSI), the American Society of Testing and Materials (ASTM), and the International Organization for Standardization (ISO). It is my opinion that companies involved in the manufacturing, production, and retailing of consumer footwear should not only be familiar with the standards that govern their industry, but apply them in the production of footwear. To suggest, as Rick had, that he was unaware of any general industry standards that pertain to the manufacturing of footwear, is shocking and reflects his company's disregard for the quality and safety of footwear bearing the fashion designer's brand.

It was my conclusion that the designer ladies' boot in question clearly presented an unreasonably dangerous risk of injury and should not have gone through the production chain unnoticed. It is my affirmed opinion that Denée had acted in a reasonable manner and was not responsible for the injuries she incurred as the result of wearing the defective designer ladies' boot.

HOW DID IT END?

The case settled for a six-figure sum. Denée recovered from her injury but has permanent nerve damage in her right foot.

24

TEED OFF!

On the morning of September 5, 2009, an elderly woman by the name of Darlene, along with her husband, Dennis, and two lady friends met at the Big Pines Golf Course located in East Texas for a round of golf. It had rained the night before, and the course had a few puddles of water. After a few holes Darlene reached the third tee box. As she got out of the golf cart, which was driven by Dennis, she saw that both the wood plank–covered pathway and adjacent grassed area leading to the tee box were wet and muddy, so she had to make a choice as to which pathway to take to get to the women's tee box. Darlene chose what she thought would be the safest route, which was the wooden-planked pathway. However, as Darlene stepped from the grass onto the wooden-planked pathway, her left foot suddenly slipped out beneath her, causing her to fall onto the elevated railroad ties. Darlene injured her right arm and shoulder.

I was retained by Darlene's attorney to provide opinions related to the safety of the golf course tee box and approach.

The wooden pathway leading from the cart path to the third tee box was made from recycled wooden railroad ties, which were heavily worn and in poor condition. Located approximately midway along the path was a single elevated step, which measured approximately four inches in height. The edges (nosing) of the step were rounded due to heavy wear. The width of the wooden pathway was approximately five feet.

INDUSTRY STANDARDS

Numerous industry standards have been developed to protect the public from unnecessary risk of harm due to a hazardous walkway. One such standard is that of the American Society of Testing and Materials (ASTM) F-1637-02 "Standard Practice for Safe Walking Surfaces," which requires the following: Section 5.1.1 entitled "Walkway Surfaces" of the ASTM F-1637-02 standard requires that "Walkways shall be stable, planar, flush, and even to the extent possible. Where walkways cannot be made flush and even, they shall conform to the requirements of 5.2 and 5.3."

Section 5.2.1 requires that "Adjoining walkway surfaces shall be made flush and fair, whenever possible and for new construction and existing facilities to the extent practicable."

Section 5.7.1, entitled "Exterior Walkway," requires that "Exterior walkways shall be maintained so as to provide safe walking conditions." Section 5.7.1.1 requires that "Exterior walkways shall be slip resistant." Section 5.7.1.2 requires that "Exterior walkway conditions that may be considered substandard and in need of repair include conditions in which the pavement is broken, depressed, raised, undermined, slippery, uneven, or cracked to the extent that pieces may be readily removed."

Section 5.7.2 requires that "Exterior walkways shall be repaired or replaced where there is an abrupt variation in elevation between surfaces. Vertical displacements in exterior walkways shall be transitioned in accordance with 5.2." Section 7.2 Short Flight Stairs (Three or Fewer Risers), Subsection 7.2.1 states that "Short flight stairs shall be avoided where possible." Section 7.2.2 states that "In situations where a short flight stair or single step transition exists or cannot be avoided, obvious visual cues shall be provided to facilitate improved step identification. Handrails, delineated nosing edges, tactile cues, warning signs, contrast in surface colors, and accent lighting are examples of some appropriate warning cues."

Furthermore, Section 4.5.1 of the State of Texas Accessibility Standards (TAS) requires that "Ground and floor surfaces along accessible routes and in accessible rooms and spaces including floors, walks, ramps, stairs, and curb ramps, shall be stable, firm, slip resistant, and shall comply with Section 4.5." This requirement that walkways be slip resistant is also required in Section 302.1 of the American National Standards Institute (ANSI) standard ICC/ANSI A117.1-1998.

It is my opinion that the defendant's pathway in question was not in compliance with the industry standards and that the wooden pathway in question had been worn as to create a series of depressions. The depressed sections allowed water to pool when it rained or when the tee box was watered. Furthermore, when the depressed section of the pathway pooled, the water could remain puddled for enough time as to create an unreasonably dangerous condition to golfers (pedestrians) accessing the third tee box.

Railroad ties, like those used at the defendant's golf course, are treated with creosote, an oil-like material used to preserve them from the elements. Such treatment will, over time, leach from the wood fibers to create an oil-like coating on the surface. The combination of water and oil/creosote residue would make for a very slippery walkway condition. According to Darlene, there were quite a few other individuals who had reportedly slipped and fallen at that same location, and she was a witness to one such event.

It is my opinion that the defendants failed to exercise reasonable care in protecting their invited guests from the previously known risk of a slip and fall and therefore were negligent in their efforts to protect Darlene. Had the defendant addressed the pedestrian walkway hazard prior to Darlene's visit, it is unlikely that she would have slipped and fallen. It was my opinion that the defendant was directly responsible for her injuries.

HOW DID IT END?

I was summoned to trial, appeared at a small East Texas courthouse, and provided testimony as it relates to the standard of care for golf courses. The jury found in favor of the defendant. Most jurors felt that Darlene approached the tee box at her own risk and that she knew that it was flooded before she got there. Furthermore, the jury felt that because Darlene was wearing a pair of low-profile spiked plastic golf shoes, which had a lower level of traction on hard surfaces, she should have exercised more caution.

25

SLIPPED ON SPONGEBOB SQUAREPANTS!

May of 2015 brought a call from a defense attorney in Oklahoma with whom I had worked on numerous cases. His firm represents a major retailer, who from time to time gets sued for a customer-related fall. The case he had involved a woman by the name of Karen, who at approximately 3:30 p.m. on May 14, 2012, claims to have slipped on a SpongeBob SquarePants night-light that was lying on the floor of one of the store's aisles. I asked if there were any witnesses to the fall, and he said there were not. About a week later I received a thick file from his office, which included the actual SpongeBob SquarePants night-light that the plaintiff allegedly slipped on. Surprisingly it was in perfect condition.

The facts of the case stated that the aisle was of a natural brown-colored polished concrete material that was clean, dry, and in good condition. The aisle was free and clear of any clutter or displays and was well lighted. The size of the SpongeBob night-light packaging measured 5.8 inches tall by 3.9 inches wide by 2 inches tall. The night-light and background packaging were both bright yellow in color. Lastly Karen was not carrying anything and was wearing a pair of flip-flops at the time of her fall.

The details of Karen's alleged slip-and-fall event are unknown, and there were no eyewitnesses. Karen stated in her deposition that she did not recall the store's layout or the exact location of her fall and that even though she recalled picking up the SpongeBob night-light after her fall, she could not recall the size or color of the night-light or the color of the floor that she slipped on. Karen further stated that the shoes that were displayed on the aisle distracted her from seeing the night-light on the floor.

It is my opinion that the polished concrete floor as used in the retail store in question is customary for a retail store and followed all state, local, and national building codes and safety standards and did not present an unreasonable risk of harm to pedestrians. The footwear aisle was well lighted via overhead lights, which were identical to those of adjacent store aisles.

Dropped merchandise, like the night-light in question, is common in the retail store industry, whereby once recognized by store employees, such merchandise is removed from the floor and restocked. The incentive to remove dropped merchandise is one that not only protects shoppers but decreases financial loss for the retailers due to reduced product damage. Karen acknowledged in her deposition that she was unaware of how long the night-light in question was on the floor prior to her stepping on it or how the night-light got on the floor in the first place. It is likely that the dropped night-light had only been on the floor but a matter of minutes before she stepped on it.

Furthermore, there is no evidence to suggest that any of the store's employees were aware of the night-light's presence on the floor prior to Karen's alleged slip and fall. There is also no evidence to suggest that there was any failure on the part of the defendant to recognize and remove the night-light in a timely manner. Given that the SpongeBob night-light was bright yellow and of substantial size, such object as it rested on a darker/contrasting-colored concrete floor would have been obvious to a reasonable person walking at a normal pace. Karen stated that she was not pushing a shopping cart at the time nor was she carrying any object, either of which may have obstructed her view of the night-light.

What I found interesting was that there was no damage to the night-light's packaging. One would expect that if stepped on, the plastic packaging would have some sign of damage, which raises the question, how is it then that Karen slipped? Although store surveillance video was working, the location where Karen allegedly fell was off camera. When combined, it was my view that SpongeBob SquarePants most likely did not cause Karen to slip and fall.

Finally, it was my opinion that the defendant was not negligent in their duty to provide a safe walking surface for their invited guests. Based upon the evidence provided to me at that time, it was my opinion that Karen's alleged slip-and-fall incident was caused by her admitted preoccupation and lack of attention to her surroundings. This point is further bolstered by her failure to recall any details of the event, the color of the floor, the specific location on the aisle she fell on, the store layout, or the actual size of the night-light she picked up and held immediately after she allegedly slipped and fell. Such failure to remember such basic information is inconsistent with the type of incident and related injuries described by Karen. Although unfortunate, it is my view that Karen's fall could have been avoided had she been paying more attention to her surroundings.

HOW DID IT END?

The retailer settled the case for a marginal/nuisance amount. Sponge-Bob SquarePants was absolved of all responsibility.

26

I HATE GRAPES!

Phyllis was an elderly woman who one day went to her local Price Crusher grocery store to purchase her weekly groceries. As she was walking down the produce aisle, she slipped and fell on grapes and grape juice that were on the floor next to the grape display. Phyllis broke her hip and seriously injured her right shoulder. The produce aisle floor was composed of a grouted ceramic tile. Caution signs were not posted on or adjacent to the walkway nor was there any form of slip-resistant carpeting, matting, or grating on the floor at the grape display.

INDUSTRY STANDARDS

As related to pedestrian safety, nationally recognized industry consensus standards serve as the basis of establishing what is "reasonable" versus what is "unreasonable." The basis by which a property owner properly and safely constructs, maintains, and inspects their walkways is defined within nationally recognized consensus standards like those published by the ASTM and ANSI. Furthermore, federal, state, and local laws, codes, and treatises are built upon nationally recognized industry standards and are frequently referenced throughout the body of law, treatises, and codes.

Section 5.1.3 of the American Society of Testing and Materials (ASTM) F-1637-10 entitled "Standard Practice for Safe Walking Surfaces" requires that "Walkway surfaces shall be slip resistant under expected environmental conditions and use." Section 5.1.4 requires that "Interior walkways that are not slip resistant when wet shall be maintained dry during periods of pedestrian use." 5.4.3 states that "Mats or runners should be provided at other wet or contaminated locations, particularly at known transitions from dry locations."

OPINIONS

It was my opinion that although ceramic tile is an appropriate walkway material for a grocery store, it does require proper maintenance and inspections. Supermarket walkways take a great deal of abuse from foot traffic, shopping carts, and exposure to accidentally spilled merchandise, which contaminate the walkway surface, causing it to become slippery. In fact, accidental spills of food products including grapes are common in the supermarket industry and are the leading cause of both employee and guest injuries. For this reason, the supermarket industry has adopted a rigorous walkway inspection or "floor sweep" procedure that requires one of the following:

1. Zone sweep, whereby employees in each department are to be on the constant lookout for floor-borne hazards such as spilled merchandise or conduct a scheduled "sweep" of their area.
2. Scheduled store sweeps, whereby a designated employee, usually a manager, performs an aisle-by-aisle inspection at a designated time of day.

Walkway inspections are the first line of defense and play an important role in preventing walkway slips and falls. Those retailers who perform a timed inspection will record the time the inspection took place via a written "sweep log." Store employees who perform such inspections require proper training as to what constitutes a floor hazard. Just because an employee walked down a particular aisle at a particular time does not constitute a proper inspection. Depending on store size and department location, floor sweeps are performed either every half hour or on an hourly basis. A daily completed log of the floor sweeps is then maintained by the store manager for a period of ninety days.

It is my understanding that the defendant performed hourly store sweeps and utilized an electronic sweep log system that required employees to perform scheduled sweeps along a preset path whereby special magnetic markers were placed at designated points along the path. Employees carried an electronic, wand-like device with which they touched each of the magnetic markers on the path. The pathway traversed by the employee was provided by the defendant; however, I only received a portion of the electronic pathway diagram, which did not actually show the produce area grocery pathway.

Furthermore, the store manager, Steve, stated in his September 26 customer incident report that "a sweep walk was conducted at 6:00 and that the Team Member who conducted the sweep had not noticed anything in the area at the time of the walk." However, when I reviewed the store's electronic sweep log, the only produce area entry recorded at or near 6:00 p.m. was one at 5:58 p.m. for produce zone 1 and another at 6:02 p.m. at produce zone 2. However, because of the incomplete electronic pathway diagram, which does not show the location and pathway for produce zones 1 and 2, I was uncertain as to the path employees traveled to reach these zones. Therefore there is no evidence to confirm that a store employee actually inspected the walkway in front of the grape display.

Based on the facts as presented to me, it was my opinion that Phyllis's grape-related slip-and-fall event could have been prevented. Price Crusher was aware of the risk of grape-related guest injuries and produced a slip-and-fall-prevention training video called "Slips, Trips, and Falls," which directly speaks to the importance of removing dropped grapes from the floor and shows a produce area employee placing carpet matting along the produce display. The video shows a female employee discussing how often grapes fall from their displays and onto the floor and states, "They are always trying to wiggle out of their displays and onto the floor, and there they are right under foot," whereby the video shows grapes falling onto the floor. The segment concludes with a Price Crusher employee placing a carpet mat along the produce aisle. Carpet matting is also recommended in the Price Crusher "Safety 24/7 Accident Prevention" to prevent slips and falls. Therefore, the defendant was not only aware of the elevated risk presented by dropped grapes but had a specific training video on the subject and in-store solution (i.e., floor matting), which the store in question consciously chose not to use.

Photographs of the Price Crusher produce aisle show grapes being sold both in clear bags as well as loose. The grape display did not have a raised or elevated front guard, which would serve to contain loose grapes that may fall out of their bags. Because of such, it would be expected that grapes as they were displayed would fall onto the floor.

Phyllis's slip-and-fall event could have been prevented had the defendant been more vigilant in their walkway inspection procedure and time cycle. Had the defendant performed floor sweeps of the grape area more frequently, provided a more slip-resistant flooring material, and/or retrofitted their merchandising/packaging of grapes as to reduce the risk of dropped merchandise, it is unlikely that Phyllis would have stepped on a loose grape, causing her to slip and fall. Unfortunately, the defendant failed in their duty to exercise good judgment and in turn, exposed their invited guests, specifically Phyllis, to unnecessary risk of a slip and fall. It is clear that had the defendant addressed the grape hazard prior to the time of Phyllis's visit, she would not have slipped and fallen and in turn injured herself.

It was therefore my opinion that the Price Crusher supermarket failed to exercise reasonable care as it relates to keeping food from creating a slip-and-fall risk and therefore was directly responsible for Phyllis's fall and related injuries.

HOW DID IT END?

Phyllis received a settlement of $109,000, which was less than her incurred medical expenses. Six weeks after receiving her settlement, Phyllis died from complications of her hip surgery. Since Phyllis was a Medicare recipient, her settlement, less her legal expenses, went to reimburse Medicare.

27

MEALS ON HEELS!

In December of 2008 I was retained by a plaintiff by the name of Josephine, who on a July evening in 2007 was a guest at a business function being hosted poolside at a hotel casino in Mississippi. As Josephine was walking from a buffet line to a dining table, she slipped on a wet floor located next to a poolside bar. The flooring material was a smooth, glazed ceramic tile and was wet from bathers' dripping swimsuits. Wet-floor signs were not posted nor were floor mats/runners installed. At the time of her fall Josephine was wearing high-heeled shoes. Josephine's slip-and-fall event resulted in her fracturing her ankle.

The facts of the case were:

1. The ceramic-tiled walkway surface provided a low level of slip resistance as compared to that of adjacent walkways, including that of the swimming pool deck and bartender work area. Such a significant level of slip resistance would be noticed by individuals entering and exiting the pool bar onto the swimming pool deck.
2. Halogen ceiling spot lighting was directly above the area where Josephine slipped and fell.
3. Two surveillance video cameras were located within the pool bar, one of which was pointed at the very location where Josephine had slipped and fallen.

The gloss level of the ceramic tile was such that it made seeing water on the floor difficult (e.g., from a swimsuit or spilled drink). Had the floor had a lower gloss level, like that of the adjacent pool deck coating, it would be easier for pedestrians to recognize the presence of water on the floor. This problem was further compounded by the fact that directly above the location where Josephine slipped and fell were bright halogen spotlights. Such lighting served to further mask the presence of water on the tile and in turn directly contributed to Josephine's slip and fall.

Both the swimming pool deck and bartender work area were coated with what appeared to be a slip-resistant coating. A cursory examination revealed that the slip resistance of the coating was significantly higher than that of the glazed ceramic tile. Why the defendant did not simply coat the pool bar area with the same material used on the pool deck and bar area is unknown; however, such material would have provided a significantly greater level of traction and in turn provided for an elevated degree of safety.

Finally, the defendant hired an expert by the name of Keith to take slip-resistance measurements of the floor. Having reviewed Keith's report, I found several errors. First, Keith used a test apparatus that is not recognized by any national authority as being an acceptable method of testing the coefficient of friction of ceramic tile. The Technical Products test device employed by Keith is notoriously inaccurate and prone to misuse by the user. Although Keith claimed to have used a device like the ASTM C-1028 device, it in fact was not. Secondly, although Keith tested the tile with three different sensor materials, rubber, leather, and silicone, none of these sensor materials are accepted per the ASTM C-1028 standard as a suitable material. The only sensor material recognized per the ASTM C-1028 standard is Neolite®, which Keith did not use. Because Keith had not used the appropriate test device nor had he used the appropriate sensor material, the subsequent data contained in his December 8, 2008, report should not be relied upon as accurate or in compliance with the ASTM C-1028 standard.

It is my opinion that multiple failures had occurred, and Josephine's slip and fall could have been prevented. First, because of the location of the accident and related flooring material, specific concern should have been given to provide a more suitable walking surface given the likelihood that the floor in question would often be wet. The defendant claimed that "the subject area contained appropriate tiles . . ."; however, a smooth-surfaced glazed ceramic tile does not provide a high degree of slip resistance when wet and therefore would be considered inappropriate for such a location.

The risk of a slip and fall is further elevated when invitees are entering and exiting the tiled floor barefooted. Had the defendant used a more appropriate flooring material that offered a higher degree of wet slip resistance (e.g., low-pile carpet), it is most likely that Josephine would not have slipped and fallen. Furthermore, it is the industry standard to utilize walk-off mats or runners in areas where water may accumulate or be migrated.

The American Society of Standards and Materials (ASTM) has published a standard entitled F-1637-02 "Standard Practice for Safe Walking Surfaces." Listed below are relevant sections of this standard.

Section 5.1.3: "Walkway surfaces shall be slip resistant under expected environmental conditions and use. . . ." 5.1.4: "Interior walkways that are not slip resistant when wet shall be maintained dry during periods of pedestrian use." 5.4.3: "Mats or runners should be provided at other wet or contaminated locations, particularly at known transitions from dry locations. Mats at building entrances also may be used to control the spread of precipitation onto floor surfaces, reducing the likelihood of the floors becoming slippery."

Section 5.4.4: "Mats shall be of sufficient design, area, and placement to control tracking of contaminants into buildings. Safe practice requires that mats be installed and maintained to avoid tracking water off the last mat onto floor surfaces."

It is my opinion that the defendant was not in compliance with the ASTM standard.

Finally, and perhaps most importantly, was the absence of appropriate hazard identification. As the front-line defense in the prevention of slip-and-fall accidents, hazard identification is an important tool. By not posting an appropriate "Wet Floor" sign, Josephine was exposed to an unrecognized hazardous condition.

Based on this information, the opinions named above served to support my initial opinions that the glazed ceramic tile was inappropriate for use in a pool bar and was unreasonably dangerous. Furthermore, the defendant had a more suitable coating material available to them that would have provided an enhanced level of traction. Had the defendant simply coated the pool bar floor in the same material as that of the adjacent pool deck, it is unlikely that Josephine would have slipped and fallen. It is my opinion that the defendant failed to protect the public from unnecessary injury and that their poolside floor did not provide a reasonable degree of care as it relates to the safe inspection and mainte-nance of their walkways.

HOW DID IT END?

Josephine received a small settlement, but it was enough to cover her medical expenses.

28

THE HIGH COST OF OIL!

In December 2004 I was retained by a plaintiff by the name of Tracy, who in the course of working for a Central Texas motor oil production plant on the quart oil filling line slipped and fell on a piece of cardboard that was on the floor adjacent to the production line. The cardboard was placed on the floor beneath the oil filling line by a company employee as a means of absorbing spilled oil from the assembly procedure. Tracy's supervisors had not inspected the walkway in question to ensure its safety. As a result of her fall, Tracy tore her ACL, which required surgery.

It is my opinion that multiple failures had occurred that could have prevented Tracy's slip-and-fall incident. Tracy had stated that in the course of her work on the oil bottle filling assembly line, she often experienced difficulties whereby the bottles to be filled did not align properly with the filling tubes, and motor oil was accidentally deposited onto the workplace floor. Such occurrence took place regularly; Tracy was trained to stop the line when bottles became misaligned. At approximately 10:30 a.m. on July 30, 2001, just such a problem had occurred on the line Tracy was operating. Tracy stated that as she was attempting to reach the turn-off button of the machine, she "stepped on a cardboard that was on the floor and fell," injuring her arm and right shoulder. This was confirmed by her supervisor, Rose, in her report dated July 30, 2001.

The use of cardboard as a floor mat is inappropriate and dangerous in a manufacturing environment. The Occupational Safety and Health

Administration (OSHA) would consider such use as a violation of its Code of Federal Regulations (CFR) Section 1910.22 Working-Walking Surfaces. Section (a)(1) of the CFR states that "All places of employment, passageways, storerooms, and service rooms shall be kept clean and orderly and in a sanitary condition." Section (a)(2) states: "The floor of every workroom shall be maintained in a clean and, so far as possible, a dry condition. When wet processes are used, drainage shall be maintained, and false floors, platforms, mats, or other dry standing places should be provided where practicable."

The use of cardboard as a walkway material would not be in compliance with the above-named OSHA requirements and would be a finable violation under OSHA's General Duty Clause. Although absorbent to liquid spills, cardboard is not a slip-resistant material and is likely to become slippery when contaminated with oil.

The use of such a mat is also a violation of the American Society of Testing and Materials (ASTM) Standard F-1637-95 Standard Practice for Safe Walking Surfaces. Section 4.4 Mats and Runners, Section 4.4.3, states that "Mats or runners should be provided at other wet or contaminated locations, particularly at known transitions from dry locations." Section 4.4.5 states that "Mats, runners, and area rugs shall be provided with safe transition from adjacent surfaces and shall be fixed in place or provided with slip resistant backing." The cardboard walkway was not in compliance with this standard.

Automated assembly lines that expose employees to slip hazards require the use of an appropriate floor matting material. Such materials offer both enhanced traction and proper drainage of liquids and are designed to prevent employee slipping and tripping. Examples of such safety matting are readily available from a wide range of industrial or safety supply distributing companies.

Finally, it was evident that the defendant was not performing routine safety inspections of the assembly line in question. Had an inspection of the floor been performed, it is unlikely that Tracy's accident would have occurred.

Rose stated such in her Supervisor's Accident Investigation Report. Section C of the report entitled "What should be done to prevent repeat of similar incident?" suggested "cleaning a spill immediately with a mop and rags not cardboard" and that "all supervisors including myself should never stop stressing the importance of being aware of where you are, what you are doing & what's going on around you." Such statement is an admission that the company's management was aware that the cardboard material was inappropriate for the application and should not have been used.

It is therefore my conclusion that the defendant had in fact created the above-named slip hazard that directly caused Tracy to slip and fall. Had the company's management addressed the above-mentioned walkway hazard, it is unlikely that she would have slipped and fallen. It is my opinion that the defendant was directly responsible for her injuries.

HOW DID IT END?

Tracy settled her claim for $92,000 and still works at the plant.

29

ROTISSERIE CHICKEN!

It was April 22, 2011, and just another afternoon in St. Louis, Missouri, when Amber and her fiancé, Shane, went to their local warehouse club store to purchase items for their upcoming wedding reception in May.

She was thirty-nine years old, Sharon and Herb's only child, and the light of their life. While shopping, Amber slipped and fell on what appeared to be spilled chicken fat from a rotisserie chicken displayed just a few steps away. The chicken was a favorite item for shoppers and came prepackaged in a clear plastic "clamshell" container, which was notorious for leaking, especially when placed into the top section of a shopping cart. You see, it doesn't quite fit, so customers have to lean it sideways. Most people don't like placing the container in the large area of their shopping cart, which is used for transporting large, heavy items. After all, just about everything sold in the store comes in "jumbo-size packs," except, of course, the rotisserie chicken.

When Amber slipped, she did the splits landing with her left leg forward and her right knee straight down, striking the polished concrete floor with the full force of her body. At first Amber thought she was all right and did not know what she had slipped on. Shane was next to her the whole time and said that it looked like smeared oil or grease. He showed her a photo that he took on his cell phone. After getting to her feet Amber knew that she was not all right and asked Shane to take her to her apartment. After a few days of icing down her knee Amber went to the emergency room, where she was diagnosed with a dislocated patella and strained medial collateral ligament. Although this type of injury is serious, it is rarely fatal; that is, unless you had previous injury to that same knee like Amber did. Amber was admitted into the hospital and scheduled for surgery to repair her knee. Amber came through the surgery just fine but then developed complications. On May 6, 2011, Amber died as a result of a "massive pulmonary embolism." She was buried on the very day she was to be married to Shane.

According to Amber's family's lawsuit, Amber (the decedent) was a business invitee on the defendant's premises when she was caused to fall by a slippery foreign substance on the floor. At all relevant times, the defendant owed the decedent a duty of personal care and safety. The defendant breached that duty of personal care and safety owed to the decedent. The defendant was negligent in one or more of the following respects:

1. Defendant caused, suffered, and permitted a slippery substance to come into contact with and remain on the floor when defendant knew, or in the exercise of reasonable care should have known, that the substance created an unreasonable risk of harm to customers.
2. Defendant failed to properly detect the substance so that it would not cause a customer to fall.
3. Defendant failed to remove the substance so that it would not cause a customer to fall.
4. Defendant failed to barricade the area covered in the substance so as to warn and avoid injury presented by the slippery substance to customers, including decedent.

I was retained by Amber's parents, who asked me to provide opinions related to the appropriateness of the store's flooring material, maintenance procedures, company safety policies, and employee safety training. I used as the basis of my opinions the nationally recognized and published consensus standards and the company's written safety policies and training procedures. I also relied upon the deposition testimony of store employees and her fiancé, Shane. I did not author a report but was deposed by three of the defendant's attorneys for nearly six hours. It was one of the longest and most grueling depositions I ever had to do.

In the end it was clear to me that the store clearly not only failed to comply with the published industry safety standards but did not follow their own policies related to guest safety. Furthermore, they were aware due to past complaints that their rotisserie chicken containers leaked. During my deposition, I stated that one way to prevent such leaks is to either bag the chicken before placing it in the plastic clamshell or use a package that doesn't leak. To my surprise the attorneys pushed back, saying, "Do you have any idea how much rotisserie chicken my client sells and the added packaging cost?" I answered, "No, I don't, but I do know the cost to Amber and her family and friends." Later, and before trial, I did a little research as to the cost to use heat-resistant food-grade bags, which to my surprise were made for rotisserie chicken. The cost ranged between two and four cents when purchased in volume. Armed with this information, I was prepared for the upcoming Monday trial. However, late Sunday night I received a call from plaintiff counsel that the case had settled.

HOW DID IT END?

Although the case was settled, it left me unsettled. I invested a lot of myself in this case, which at the time garnished a lot of local media attention. Although I never met Amber's parents, as a parent myself, I grieved for their loss. For a parent to lose their only child for what was something so stupid and easily correctable is very difficult to deal with. As a tribute to Amber and her family, I requested that the board of directors of the National Floor Safety Institute (NFSI) agree to include the name of a real victim on all of the ANSI standards authored by the NFSI B101 committee on slip, trip, and fall prevention. The board approved the suggestion by a unanimous vote. The first ANSI standard published by the committee was the B101.1 standard, which is now dedicated in the memory of Amber. Amber was not just a slip-and-fall statistic but a person. A person who loved life and will forever be missed by those who loved her.

30

UNLUCKY AT THE CASINO!

In 2015 I was retained by a plaintiff attorney representing the estate of a gentleman by the name of Wally, who on March 3, 2014, slipped and fell on the exterior parking lot sidewalk of the Gold Mine Hotel and Casino located in Mississippi. According to the information collected, it appears that the night before brought a terrible ice storm to the small town where the Gold Mine Hotel and Casino is located. Shortly after midnight the power in the hotel and casino went out, whereby only the casino remained open and heated. As a precaution, Wally went out to remove a few items from his car to prevent them from freezing. He walked out of the building toward his car, which was parked near the handicap parking lot. The sidewalk outside of the building was free and clear of ice and had ample deicing compound sprinkled on it; however, as Wally continued walking along the hotel's sidewalk toward his car, there was no deicing compound. Just a few steps away from his car, Wally slipped and fell, fracturing the back of his skull. His body was found frozen to the sidewalk the following morning by a couple walking their dog. Paramedics were called to the scene and could only get to Wally's body by crawling on their hands and knees.

The facts were:

1. Wally slipped and fell on the exterior disabled parking lot walk-way located on the defendant's property. The walkway in question was a main entrance into the hotel and casino.

2. On the day of Wally's visit to the defendant's property, the exterior walkway on which he slipped and fell was covered in a layer of heavy ice from the previous night's ice storm.
3. Wally struck the back of his head on the iced sidewalk consistent with a violent slip-and-fall event.
4. The accumulated ice was not properly removed by way of salt, sand, or deicing compound to create a uniformly safe and slip-resistant walking surface.
5. Neither warning signs, barricades, nor any other warning methods were used by the defendant to alert the public, specifically Wally, as to the pending ice hazard.

It is my further understanding that on the morning of March 3, 2014, the defendant was aware of the icy walkway conditions and therefore was on notice as to the dangerous condition on their premises. The defendant was further on notice of the fact that inclement weather was approaching and had ample time to take safety measures to protect its guests. Numerous witnesses as well as several of the defendant's employees who were on the premises on the day of Wally's slip-and-fall event confirmed that the exterior walkway was icy and presented a slip hazard.

The hotel guest who found Wally was a middle-aged woman by the name of Heidi who stated in her deposition that the condition of the sidewalk where Wally slipped and fell was "thick ice" that had "glazed over" and that the adjacent grass was "wet, crunchy." To get to Wally she had to crawl and "scooted sideways on her hands and knees" to reach him as he lay on the sidewalk. Heidi further stated that her husband, Mike, told her that he had slipped and fallen moments before she exited the hotel side entrance. Heidi further stated that as she returned to the hotel after performing CPR on Wally, she did not see any sand, salt, or deicing agent on the sidewalk, nor did she see any "warning cones or areas that were roped off or anyone standing around from the casino or hotel telling people to be careful because of the glaze of ice on the sidewalk or patio." Heidi further stated that she did not see anyone from the hotel or casino spreading any salt, sand, or deicing agent, but once she returned to the hotel lobby, she recalled hearing other guests saying that "there were employees throwing out salt on the area."

Heidi's husband, Mike, stated in his deposition that upon leaving the hotel's side entrance he slipped and fell on the ice-covered sidewalk, which he identified as being "glisteney." He went back into the hotel lobby to get his wife. To ensure that they did not slip and fall on the exterior walkway, he and his wife got down on their knees and "crawled along the edge of the sidewalk down—or edge of the grass down to where he [Wally] was and immediately began CPR." Mike stated that the location where Wally was lying on the sidewalk had "more ice and snow" than the area of pavement closer to the building and adjacent to the building's entrance. Mike said that he witnessed an EMT slip and almost fall, but the EMT "caught himself with his hand before he got down." Lastly, Mike stated that he did not see any salt or icemelt on the walkway where Wally slipped and fell, but later that afternoon he recalled seeing salt or icemelt on that section of the walkway.

Drew was an engineer employed by the defendant and stated in his deposition that the defendant did not have any written policies related to what he was to do when there was inclement weather, nor was he "provided any direction" related to ice removal. Drew stated that he was aware that the defendant had "big pallets of salt" on hand to use to deice their sidewalks and drives. Although there was plenty of salt on hand, Drew stated that the defendant did not have "a set rule" as to "where you're going to start and where you are to finish" spreading the salt.

Brian was an engineer employed by the defendant and stated in his deposition that as it relates to ice formation on the defendant's property, "there should be no ice out there," and that the defendant had an ample supply of salt including "bags of rock salt."

Dave was an engineer employed by the defendant and stated in his deposition that the defendant "knew ice was coming," that there was "a fine mist that had not started to freeze," and that "they were prepared for it to start [to freeze]." He further stated that "every time it ices he takes steps to clear the entrances," including "anywhere he expects customers to walk."

John was an engineer employed by the defendant and stated in his deposition that spreading salt, sand, and deicing agent is the most important when there is inclement weather "because if the guests can't get in safely, the employees of the Gold Mine Casino can't guarantee that they have a good time." He further stated that as related to spreading sand, salt, or icemelt, "those are the first things we do once snow hits the ground, because we have to make sure that everybody makes it in safely," that "the sidewalks and entrances come first," and that the "entrances and sidewalks are first priority." He further stated that "it's our job to make sure that no one gets hurt, which is why we spread salt down on the ice and snow, yes sir." John also stated that guests are "on their own" and "responsible for their own safety" and that when they are walking on icy sidewalks they "go out at their own risk" and that "they are on their own once you walk out there."

Zach was the director of safety and risk manager for the property in question and stated in his deposition that Clay had kept him informed as to what was happening on the property at the time of the ice storm and power outage and that Clay had "taken care of the entrances. Because of the weather conditions, some of the areas they were unable to deal with because of the melting and then refreezing" and that "the icemelt they applied was not working as it should." Zach stated that it was his understanding that Clay was "doing everything possible to make it as safe as he could." Zach also stated that he was on-site on or about the time Wally was being transported by the paramedics, and as he walked down the pathway toward Wally, he walked over snow and ice that was on the ground. But he stated that he did not slip and that "there was not ice on the ground," a direct contradiction to other eyewitnesses as well as the depiction in the short videos and photos taken of the scene. Zach stated that he did not see any evidence of ice treatment and that he "didn't see any particles" on the walkway but understood that because Clay told him that he had treated the areas around the entranceways, he believed they were treated. Zach also stated that "under the conditions they were facing that day it was the best they could do and their crews were working diligently to make it as safe as we possibly could" and that "in an ideal world, under ideal conditions, sure, we would like to have the sidewalk cleared." Even though Zach was aware that the location where Wally slipped and fell was a walkway leading to the disabled parking lot, he would expect Clay and his staff to "prioritize and take the most serious areas that affect our customers the most and deal with those." Zach later contradicted himself by stating that the disabled parking lot sidewalk was in fact hazardous and that it "was icy" and was "a danger" but that "in this particular incident, I don't think it was neglected. I think it was, you know, it was attempted to be—to be cleared."

Bob was a security guard at the defendant's casino and hotel at the time of Wally's slip-and-fall event and stated that there was ice on the sidewalk where Wally slipped and fell and that he took photographs of the location but later deleted them.

Clay was the director of facilities of the defendant's hotel and casino at the time of Wally's slip-and-fall event and stated that up to the point where Wally slipped and fell, "the sidewalk was safe." Clay said that all the sidewalks were "sprayed the night before" but that "some areas were clearer due to heat coming from the building" and that he "did the best he could" to remove the ice. Clay also stated that although attempts were made to remove as much ice from the entrances, he was unable to remove the ice farther from the building, specifically where Wally slipped and fell. The ice there was "frozen hard," but "they did as good as they could do with the manpower and staff they had to work with. . . ." Clay stated that he walked up to Wally and did not see any salt or deicing material on the sidewalk. Finally, Clay said that he was not familiar with the various temperature ratings for deicing materials.

INDUSTRY STANDARDS

Subsection 5.1.3 of the American Society of Testing and Materials (ASTM) F-1637-12 Standard Practice for Safe Walking Surfaces requires that "Walkway surfaces shall be slip resistant under expected environmental conditions and use." Section 5.7 "Exterior Walkways," Subsection 5.7.1 requires that "Exterior walkways shall be maintained so as to provide safe walking conditions." Subsection 5.7.1.1 requires that "Exterior walkways shall be slip resistant." Subsection 5.7.1.2 requires that "Exterior walkway conditions that may be considered Substandard and in need of repair include conditions in which the pavement is broken, depressed, raised, undermined, slippery, uneven, or cracked to the extent that pieces may be readily removed."

Section 9 "Warnings/Barricades," Subsection 9.1 "General" of the American National Standards Institute (ANSI) A-1264.2-2012 Standard for the Provision of Slip Resistance on Walking/Working Surfaces states that "a warning shall be provided when a slip or trip hazard has been identified until appropriate corrections can be made or the area barricaded." Section 9.1.1 "Alternate Route" states that "When a slip/fall hazard covers an entire walkway, making it difficult to safely route personnel around the hazard, barricades shall be used to limit access. (See Section 10.1.) If appropriate, assign a person(s) to detour pedestrians in conjunction with the appropriate use of warning signs until the barricade can be erected or the hazard removed."

Section 11.3 "Snow and Ice Removal" requires that "Snow removal shall be performed as expeditiously as conditions and resources allow, following a written action plan."

Section 11.3.1 "Snow and Ice in Pedestrian Walkways" states that "where snow and ice exists in pedestrian walkways, safe maintenance techniques shall include plowing, shoveling, deicing, salting, or ice-melting chemicals and sanding as foreseeable."

Section 11.3.2 "Responsibilities" states that property owners are to "establish the responsibilities for facility managers, custodians, grounds maintenance staff and contracted snow removal personnel."

OPINIONS

It is my opinion that the defendant was not in compliance with the above-named nationally recognized industry standards that establish the standard of care for sidewalk/parking lot maintenance. It is also my opinion that multiple failures on the part of the defendant had occurred that, but for the failures, could have prevented Wally's slip-and-fall event. First, the defendant's management and employees were aware of the weather forecast and likelihood of the accumulation of ice due to an approaching ice storm. Instead of being proactive and treating the sidewalks with deicing agent, they waited until the storm was upon them. After ice began to accumulate on the walkways, they made the decision to focus their clearing efforts on the entrances and exits, and even failed to do that task well. The defendant's employees all stated that they recognized that icy walking surfaces constituted a hazard. Given such, it is my opinion that the defendant failed to properly and timely remove the sidewalk ice hazard, which in my opinion was the cause of Wally's fall and subsequent injury.

Based upon my training and experience and review of the scene photos, the cell phone video, and the witnesses' description of the condition of the concrete walkway leading from the building to where Wally was found lying, it is my opinion that more likely than not, the defendant's employees had undertaken at some point prior to Wally's fall to deice the first sixty-two feet of walkway from the building to the parking lot. This would account for the differing descriptions of the condition of the walkway—"glazed" versus "thick ice." It is likewise my opinion, based upon my training and experience and review of the evidence in this case, that the defendant's efforts to clear the walkway abruptly ended short of the area where Wally fell and that the remaining thirty feet of walkway could continue to accumulate ice without any effort whatsoever to clear it. Wally was apparently able to safely egress from the hotel exit and down the sixty-two-foot-long distance, only to fall approximately twelve to sixteen feet further down the walkway. Since this walkway was the only access to the handicap parking spaces in the parking lot, the defendant should have been even more diligent in its efforts to continue deicing the entirety of the sidewalk. Given such, Wally's slip-and-fall event was the direct result of the defendant's failure to deice their walkways, specifically a walkway that they knew was in regular use by their patrons.

The defendant failed to warn Wally as to any impending slip hazard either verbally or via posted warning signs, or via a written notice to their guests. The defendant also failed to barricade the hazardous area until it could be cleared and/or made safe to walk on. Had the defendants simply finished the job by applying salt, sand, or a deicing agent to the remainder of the disabled parking lot sidewalk, this slip-and-fall incident would have been avoided.

It is also my opinion that the defendant knew or should have known that the pedestrian routes their guests would use to exit and enter their hotel on March 3, 2014, had become dangerously slippery due to the accumulation of ice that formed as a result of the previous evening's ice storm. Although the defendant's employees named above allegedly attempted to remove the accumulated ice around the entrances and exits by spreading deicing agent, they failed to do so properly as to render the walkways known to be used by its patrons uniformly safe. Deposition testimony from the defendant's employees affirmed that the defendant had the materials, manpower, and opportunity to spread deicing agents onto their walkways but consciously chose not to do so. In my opinion, their decision created an unreasonably dangerous condition to exist on their premises. The defendant's claim that the spreading of deicing agent after the precipitation had already begun would have been useless (or even made the walkway more dangerous) is completely without merit and violates common sense. The deicing product available for use by defendant was designed to melt snow and ice in temperatures ranging from 0 degrees Fahrenheit to –10 degrees Fahrenheit. Had the deicing agent available to the defendant been timely used on the sidewalk where Wally fell, it would have been free of ice and safe to walk upon.

The defendant owed a duty to their invited guests, specifically Wally, to provide a safe walking surface while on their property and to maintain the premises in a manner that made it safe for invited guests to enter and exit their buildings and property. Furthermore, in the event that a walkway hazard is recognized, the defendant has the duty to warn their guests as to any unsafe conditions on the premises. The defendant failed to follow the generally accepted safety standards referenced herein, as well as their own policies and procedures relating to ensuring customer safety by routinely inspecting sidewalks and walkways and correcting known hazards on their premises. It is my opinion that the defendant breached these duties and that such failure was the proximate cause of Wally's fall and subsequent injuries.

Had the defendants: (1) timely and properly applied the correct amount and type of deicing compound equally along the entire surface of the parking lot walkways and/or (2) removed the accumulated build-up of ice by way of chipping and shoveling and (3) properly posted warning signs alerting their invited guests as to the impending slip hazard or (4) notified their guests in person or in writing as to the hazardous condition of their exterior walkways and request that they do not venture outside and/or (5) barricaded the untreated icy section of sidewalk as to prohibit entry, more likely than not, Wally would not have slipped and fallen and suffered injury. Furthermore, the fact that the defendant did not have an inclement weather policy that provided training to its employees regarding what to do when there is inclement weather and direction as it relates to ice removal, is further evidence that removing ice from pedestrian walkways was not a safety priority.

It is my concluding opinion that the defendants were negligent in failing to provide reasonable care as it relates to providing a safe walking path and/or entrance path to their invited guests, specifically Wally. Although the defendant's employees felt that they did the best they could, they did not. They utterly failed to clear a walkway that should have been cleared. They could have done better by simply following the above-named steps. Had they done so, it was my concluding opinion that Wally would not have been exposed to the serious slip hazard that led to his slip-and-fall event and blunt-force trauma to his head.

HOW DID IT END?

The case never went to trial but was settled for a seven-figure sum.

31

BROKEN TOILET!

On July 10, 2007, Jaime, along with his family, checked into room number 207 of the Hamilton Inn Hotel in Wichita, Kansas. It was the night before his son's baseball tournament and everyone wanted to get a good night's sleep. Rather than wait until morning, Jaime decided to take a shower that night. A few minutes into his shower, Jaime suddenly and unexpectedly slipped in the shower/bathtub, falling backward, causing his body to exit the bathtub and his head to strike the top of the bathroom's commode, breaking the commode in two pieces. Jaime was transported to a local hospital, where he was diagnosed with a broken neck. A few days later Jaime was told that he would never walk again and that he was a quadriplegic. Jaime sued both the Hamilton Inn Hotel and the manufacturer of the bathtub, the Coker Company, for negligence.

The defendant, Hamilton Inn, claimed that their housekeepers used a "pink spray/cleaner" and a "cream cleanser" to clean their guestroom bathtubs. The pink spray/cleaner is identified as a Disinfecting Heavy Duty Acid Bathroom Cleaner, which is manufactured by the Acme Corporation. The cream cleanser is manufactured by the Cremco Company.

One of the Hamilton Inn's housekeepers, Rosie, stated in her deposition that she was aware of another person who had slipped and fallen in one of the hotel's bathtubs prior to Jaime. "If the tub gets dark in the bottom, that's when you use the green scratchy pad?"

Maria was also a housekeeper at the defendant's hotel and stated in her deposition that "To clean the tub—to clean the tub we use a red liquid. We spray it first and then we scrub it, we rinse it with hot water, and then we dry it very well until it's dry." Maria also stated that she, like Rosie, used a "green pad" to scrub the bottom of the bathtub and that she would scrub the tub for "about five minutes," and she would then rinse and dry it.

Another housekeeper by the name of Crystal stated that she was trained by a representative from the Acme Corporation as to the use of their products, including the pink spray/cleaner, and that the manufacturer's representative's "emphasis of the meeting was to emphasize the importance of rinsing the substance (pink spray cleaner) off because it will make it (the tub) slippery."

Elizabeth was also a housekeeper at the Hamilton Inn and stated that she "rarely used the pink stuff" but almost always used the cream cleanser and scratchy pad. She followed such procedure because she "didn't like spraying and wiping the tub because she then had to rinse it out." Elizabeth also said that "a lot" of housekeepers would spray the pink cleaner on the tubs and wipe it off without rinsing, that it's "quicker not to rinse," and that other housekeepers did the same. Elizabeth also said that she was aware of other guests complaining about the bathtubs being slippery and that, in her opinion, they needed a bath mat. Anna, like the other housekeepers, did not report her recommendation to her supervisors and stated that she "didn't do anything about it or nothing because she was just—she said, I'm just making a comment."

Guests at the hotel were also deposed. A woman by the name of Chantel stated in her deposition that she had noticed that the bathtub in their room was very slippery and that she, along with her husband and mother-in-law, reported such to the hotel's front-desk clerk. Chantel said that her mother-in-law, who also was a guest of the hotel and had the same problem with her room's bathtub, stated that "she couldn't like stay in one spot, she kept slipping down." She had placed a towel in the tub to keep from slipping when she took a bath.

John, who also was a guest of the defendant's hotel at the time of Jaime's slip and fall, stated in his deposition that he had gone down to the front desk of the hotel to request a bath mat to put down since it (the bathtub) was so slippery and was told by the front desk attendant that "they don't use rubber mats anymore and to just use a towel."

Matt was also a guest of the defendant's hotel at the same time Jaime was a guest and stated in his deposition that he "noted right off the bat" that the bathtub in his room was "slick" and associated the problem to soft water; "everything was slick," including "the whole bathtub, side, and bottom." Matt was so concerned for the safety of his family that he "told his kids to take baths from now on."

A Coker Company representative by the name of Jeff stated in his deposition that his company applied the Super Safe pattern on the bottom of their tubs "to keep people from slipping." The new Super Safe pattern was "similar to what most people would call sandblasting." Jeff also stated that he had gone to the defendant's hotel and did not conduct any formal testing other than that of "an eyeball of the tub." Jeff confirmed that his company's recommended cleaning instructions prohibited the use of high-alkaline or acidic cleaners, nor did they recommend the use of "abrasive cleaners"; they only recommended the use of cleaners whose pH is between a "3 and 8" and the use of a "soft nylon brush" "once a month." Jeff stated that "strong acids or bases attack the porcelain enamel, so it's important to warn against their use" and that a high-alkaline or strong acid cleaner could make the tub in question slippery by "leveling out the peaks and valleys" or "removing and smoothing out the surface." Jeff further agreed that it was his company's recommendation to prohibit the use of abrasive pads on their bathtubs.

Jeff's company had never conducted any testing that correlated the relationship between strong acidic cleaners and their long-term effects on slip resistance. His company would "do audit testing" of their bathtubs' slip resistance, and they had them certified for slip resistance by "third parties" of which he named two. He further stated that "at the time of manufacture of this bathtub it required that the coefficient of friction be greater than 0.04 according to the test for the length of the manufacturer's warranty"; their warranty is only one year. A value of 0.04 is approximately equivalent to that of wet glass.

Jeff also stated that since 1994, his company did not conduct any prolonged testing of cleaning products that caused a bathtub to become dangerous and that he did not know if the Acme Heavy Duty Acid Bathroom Cleaner was consistent or inconsistent with Coker's instructions. Jeff was unaware of the effects that different soaps or shampoos might have on the slip resistance of their bathtubs and was not certain as to the age or level of usage at which their tubs' slip resistance would lose its effect. He "would not be exactly sure how to test for it," and "anecdotally that's not the experience we've had with this product since we introduced it in 1994."

Jeff was aware that "from a consumer standpoint," a slip and fall in a bathtub would be "the primary risk in the bathroom" and confirmed that his company did not test every bathtub they produced and that he had no data as to the actual slip resistance of the bathtub in question; there was a range of variation between each tub they manufactured, and they rely upon the use of the "statistical distribution" data they derive from their audit. Jeff believed that the concept that a bathtub is "safe is a relative term that includes experience in the user, familiarity with the product" and that "it has a physical difference in the coefficient of friction."

Jeff also stated that the slip resistance of Coker's bathtubs can be impacted by the products used to clean them and that his company provided cleaning instructions, in part, so their customers would not reduce the coefficient of friction as to cause "an individual incident within a bathtub." Jeff related the increase in the coefficient of friction of a bathtub as to make the tub "potentially less likely to slip under equal conditions," which "is one of the factors in the safety of the environment of the bathroom." He further stated that it is important that their customers follow their cleaning instructions because it provides for the maintenance of the tubs' coefficient of friction.

It was my opinion that multiple failures had occurred that could have prevented Jaime's slip and fall. First, the Hamilton Inn's management exercised poor judgment in not providing the appropriate number of grab bars. Secondly, the bathtub surface was not slip resistant and was not in compliance with the industry standards for slip resistance. Thirdly, a rubber bath mat was not provided, which could have rendered the tubs' surface slip resistant.

It is my opinion that the bathtub in question was not in compliance with the industry standards for slip resistance as established by the American Society of Mechanical Engineers (ASME), or the American Society of Testing and Materials (ASTM). The ASME standard number A112.19.4M-1994 (supplement 1-1998) states in Section 3.3.6 Slip Resistance, "The bathing surface of a bathtub shall be treated in such a manner that it shall comply with the ASTM F462. Treatment shall start

2 in. measured from all side and wall radii and 3 in. measured from the centerline of the drain and from the compound corner radii."

Based on the photographs of the bathtub, it appeared to me that the surface of the bathtub was not manufactured in compliance with this standard. Furthermore, the embedded circular pattern was discolored as to indicate soil buildup. The manufacturer applies these discs onto the surface of the bathtub for the specific purpose of improving slip resistance. However, over time, the raised circular pattern loses its slip-resistant characteristics due to wear, surface contamination, and maintenance. Such contamination will "fill in" or coat the textured discs and, in turn, negate the slip-resistant quality of the bathtubs' surface, thus increasing the risk that the bather will experience a foot slip. Such deterioration is a combination of prolonged use of soaps, shampoos, and so on, as well as the long-term effects of improper maintenance.

Section 5.1.1 of the American Society of Testing and Materials (ASTM) Standard F-446-85 (1999) entitled "Standard Consumer Safety Specification for Grab Bars and Accessories Installed in the Bathing Area" requires that "A horizontal grab bar or bars shall be installed on the back-wall critical support area with a total minimum length equal to 30% of the horizontal length of the critical support area." The specification as established by Hamilton Inn Hotels calls for a single grab bar to be mounted vertically on the back wall of the bathtub and not horizontally as required by the above-named standard and is therefore not in compliance with the ASTM F-446-85 standard.

Grab bars provide the bather the opportunity to stabilize themselves as they stand or shuffle on a slippery bathing surface. Without grab bars, the bather would be unable to adequately arrest or stabilize themselves if a foot slip was to occur. In a 2003 American Hotel and Lodging Association membership poll regarding the use of bathtub grab bars, it was found that most respondents had been installing grab bars in their hotels' bathtubs and found such practice beneficial in preventing guest-related slips and falls. The use of multiple grab bars has become the industry standard for the hotel industry. The tub in question did not have a horizontally mounted grab bar on the back wall and therefore was not in compliance with the industry standard. The use of grab bars is an inexpensive means by which to assist anyone who may experience a slippery bathing surface. The defendant did not provide both vertical and horizontal grab bars as required in the above-named standards, and therefore the bathtub/shower was not in compliance with the industry safety consensus standards.

The hotel knew their guestroom bathtubs were slippery and posed a slip hazard to their invited guests. Numerous guests as well as members of the hotel's housekeeping staff had notified their management as to the slippery condition of their room's bathtub, and the hotel's management chose not to respond nor provide a remedy. It is also clear, based on the deposition testimony of the hotel's housekeeping staff, that they were not cleaning the bathtubs properly; because of this, they were creating an unreasonably dangerous condition every time they cleaned the tubs. Based on the supplemental information listed above, the housekeepers were:

a. Using acid-based cleaning products that according to the bathtub manufacturer would harm the enamel surface of the bathtub and compromise its safety.
b. Use of an abrasive cream cleaning agent that according to the bathtub manufacturer would harm the enamel surface of the bathtub and compromise its safety.
c. Improperly applying said products via the use of an abrasive "scratchy" pad, which the bathtub manufacturer strictly prohibited.
d. Improperly cleaning guest bathtubs whereby they did not rinse the bathtubs after application of the "pink spray/cleaner."

Furthermore, based on the deposition testimony of the Coker Company's representative, they were negligent in not ensuring the safe production and maintenance of their bathtubs. Key points of Jeff's testimony included such facts as:

a. The only independent testing or "certification" of Coker's line of bathtubs was done back in August of 1992, which was two years before Coker had changed their Super Safe pattern from that of an abrasive topical coating to that of a sandblasted surface. Furthermore, given that the independent testing lab's report was more than eighteen years old, such a testing and reporting cycle would not be in compliance with any quality control industry standard, and therefore such outdated test results would not be considered a reasonable sampling. This, along with the fact that the only test report provided by the Coker Company that shows that they tested the coefficient of friction of the Super Safe surface was from June of 1995. All other coefficient of friction test reports were product specific and conducted as to demonstrate the effects individual chemical products would have on the Super Safe pattern. Given such, it appears that the Coker Company did not routinely test their bathtubs' slip resistance, and Jeff's claim that the Super Safe patterned surface has been "certified for slip resistance" was misleading and inaccurate.

b. Most test reports produced by the Coker Company were for cleaning/appearance and addressed such issues as chemical resistance and stain and soap scum removal and did not address coefficient of friction tests. Of those reports where the coefficient of friction of the Super Safe surface was the focus of testing, most of the reports referenced coefficient of friction test results; however, such data was missing from the reports.

c. The Coker Company's recommendations as to proper cleaning chemicals were contradictory as to mislead or confuse their customers. Examples include the Coker Company's prohibited use of scrubbing pads and abrasive cleansers on their Super Safe bathing surface. However, the Coker Company had recommended the use of the GOR 1 and GOR 3 cleaning agents. The GOR 1 product is an abrasive cleaner, and according to the manufacturer's instructions, requires the use of an abrasive scrubbing pad, which they provided. Interestingly, the test report, which Coker conducted as to the effects of the specific green pad/sponge as used by the Hamilton Inn's housekeepers on the Super Safe surface, was incomplete and did not contain any coefficient of friction data.

d. The Coker Company openly endorsed the use of the GOR 3 Non-Acid Bathtub Cleaner and claimed that such product would maintain the slip resistance of their Super Safe bathtub surface. However, their internal test report for the GOR 3 product clearly demonstrated a continuous and steady reduction in coefficient of friction after accelerated use, thus suggesting that the GOR 3 product would actually decrease the safety of their Super Safe surface over time, therefore exposing their customers who would use such product to the elevated risk of a slip and fall.

e. Although Jeff claimed that he was unfamiliar with the effects the Acme acidic tub cleaner had on the Super Safe surface, his company actually did test the Acme product. The report is entitled "Acme Cleaners for Coker Super Safe on Enamel Bathtubs." The Coker Company tested the Acme product via the use of a "nonscratch abrasive scrubber," which revealed an increase and not a decrease in coefficient of friction over continual scrubbing cycles.

f. Jeff stated that he was unaware of any specific chemical cleaners or treatments that could increase the slip resistance of Coker's bathtubs; however, there is evidence to show that the Coker Company was aware of at least one such product as early as 1998. The Coker Company did not produce an in-house report that I had in my possession, which revealed that they had in fact tested a product called Traction Plus Bath Traction Treatment, which according to Coker's report revealed that when the Traction Plus product was applied to their Super Safe surface, the coefficient of friction of both laboratory and in situ bathtubs increased significantly. In fact, the Coker Corporation's then director of research and development tested the Traction Plus product on his own personal bathtub and commented that the product "completely removed all dirt residue from the tub bottom" and provided detailed test data, which revealed a significant increase in coefficient of friction. The Coker test report for the Traction Plus product is dated October 26, 1998.

g. The Coker Company did not test the slip resistance of each bathtub but rather used an unreliable and unorthodox statistical data modeling method whereby a very few number of bathtubs were tested and the resultant data extrapolated to untested bathtubs. Examples of such are revealed in Coker's test reports, which were provided as responses to the plaintiff's Second Request for Production. In those reports it was revealed that when the slip resistance of the Super Safe surface was tested, the initial coefficient of friction values ranged dramatically. For example, the report showed that the original coefficient of friction ranged between a low of 0.44 and a high of 0.55, while a second report revealed coefficient of friction values ranging from a low of 0.19 to a high of 0.36. A third report for the GOR 3 showed Super Safe coefficient of friction values ranging from a low of 0.24 to a high of 0.30

and for the Traction Plus product, the coefficient of friction values ranged from a low of 0.5 to a high of 0.8. Such ranges clearly demonstrate the wide range of coefficient of friction values actually produced by the Coker Company, which cannot be used or analyzed via a statistical model.

It was my opinion that the management of the hotel were negligent in not providing a reasonably safe bathing surface to their invited guests, specifically, Jaime, and in doing so, failed in their responsibility to protect their invited guests from unnecessary risk of a slip and fall. The defendant, Hamilton Inn, could have easily made the bathtub in Jaime's room safe by simply cleaning it correctly or by providing a rubber bath mat in the bathroom.

It is also my opinion that the defendant, the Coker Company, was negligent in not manufacturing the bathtub in compliance with the industry standards listed above. Had the bathtub in question been properly manufactured as to have a level of slip resistance that did not change over time and informed their consumers as to the proper maintenance methods, it is unlikely that Jaime would have slipped and fallen and, in turn, injured himself.

HOW DID IT END?

Jaime settled with both the Hamilton Inn and the Coker Company for an undisclosed sum. Jaime is doing well but has lost a big part of his quality of life. Who would believe that something as common as a wet bathtub could change a person's life so dramatically? In 2016 the ASTM F-462 bathtub standard was withdrawn without replacement.

32

MARY AND GINGER AT THE CASINO!

On a rainy August 7, 2011, a young mother by the name of Mary, along with her daughter Ginger, were guests at the Diamond Mine Casino in Atlantic City, New Jersey. After a day of gaming, Mary and Ginger left the casino and returned to their car, which was parked in the casino's covered parking garage. As Mary approached her vehicle, she slipped and fell on a painted section of the walkway, which was wet from a slow, drizzly mist that had blown in from the open walls of the parking garage. Having reviewed the casino's surveillance video, Mary is seen walking slowly as she approaches her vehicle. At the time of her fall Mary was wearing a pair of rubber-soled sandals, which were in good condition, free of defects, and due to the rubber outsoles, provided a reasonably high level of slip resistance.

On Monday, December 10, 2012, I had conducted a site inspection of the defendant's parking garage and had taken a series of photographs. The walking surface of the parking stall where Mary had parked had been painted numerous times in at least five different colors, including yellow (two shades), black, gray, and blue, for each of which all but the original layer of paint was applied over previous layers. At least one older layer of paint appeared to contain an abrasive element to enhance the paint's traction, while later coats did not. The section of walkway where Mary slipped and fell displayed signs of heavy wear and soiling. Furthermore, I had observed numerous sections of the walkway where the paint had flaked off due to improper subfloor preparation and other areas where paint was applied over chewing gum.

Section 5.1.3 of the American Society of Testing and Materials (ASTM) F-1637-09 "Standard Practice for Safe Walking Surfaces" requires that "Walkway surfaces shall be slip resistant under expected environmental conditions and use. Painted walkways shall contain an abrasive additive, cross cut grooving, texturing, or other appropriate means to render the surface slip resistant where wet conditions may be reasonably foreseeable."

Section 5.7.1 requires that "Exterior walkways shall be maintained so as to provide safe walking conditions." Section 5.7.1.1 requires that "Exterior walkways shall be slip resistant." Section 5.7.1.2 states that "Exterior walkway conditions that may be considered substandard and in need of repair include conditions in which the pavement is broken, depressed, raised, undermined, slippery, uneven, or cracked to the extent that pieces may be readily removed."

Section 4.5.1 of the Americans with Disabilities Act (ADA) entitled "General" requires that "Ground and floor surfaces along accessible routes and in accessible rooms and spaces including floors, walks, ramps, stairs, and curb ramps, shall be stable, firm, slip-resistant, and shall comply with 4.5."

Section 10.1 "Safely Maintained" of the ANSI A1264.2-2006 "Provision of Slip Resistance on Walking/Working Surfaces": "Walking surfaces for use in accordance with 2.3 shall be safely maintained." Section 10.2 Practical Considerations: "Where it is not practical to replace flooring, etching, scoring, grooving, brushing, appliqués, coatings and other such techniques shall be used to provide acceptable slip resistance under foreseeable conditions." E10.2: "Surfacing applications and/or treatments are available that can impart increased slip resistance to problem surfaces. Some flooring surfaces can have their surface traction enhanced by etching. Certain paint on or trowel on applications can enhance slip resistance. It is important to select one that will adhere tenaciously to the substrate. Clean ability and durability should be considered. Patch testing of prospective materials in a problem environment is recommended before proceeding with general application. Carpeting is also an option worthy of consideration for control of slips. Broom finishing of concrete allows for an increase in surface projections for a rough surface with excellent frictional properties. Resilient flooring can be made less slippery with the application of a chemical coating containing abrasive granules. Other smooth, hard surface flooring such as concrete and metal can be treated with paint, trowel, or other finishes which impart roughness. Chemical resistance and durability should be considered in these applications. Some slip resistant flooring materials are manufactured with a granular surface and may not require additional treatment to become slip resistant."

Liberty Mutual Insurance recommendations for use of aggregate in exterior painted walkways describes the importance of using an appropriate aggregate material on painted walkways, whereby the defendant should have used a coarse-size aggregate in the paint as to render it slip resistant.

At the time of Mary's slip-and-fall event two security officers by the name of Ray and Joe were designated to be the "outside rovers" at the casino. Ray stated in his deposition that he allots twenty minutes of his hourly tour to cover the complete seven floors of the garage, and Joe testified that he allotted up to forty-five minutes of his hourly tour to cover the same seven floors of the garage. Ray also said that in the course of conducting his hourly tour that "we don't look for details." Such failure to check in detail would therefore make it very difficult for both Ray and Joe to recognize a walkway hazard, which can only be recognized via a more detailed level of inspection.

OPINIONS

It was my opinion that the painted walkway in question was not in compliance with the applicable industry standards, and because of such, represented an unreasonably dangerous condition. The walkway in question was improperly coated and maintained as to provide an unreasonably dangerous condition. Specific walkway defects include (1) the top layer of paint did not contain an aggregate material as to render the surface slip resistant when wet; (2) the paint was improperly applied as to not adhere to previous layers of paint; and (3) underlying layers of paint should have been removed by mechanical shot blasting or other appropriate means as to provide both adhesion and surface texture to enhance the surface's slip resistance. Given the parking garage's openness to the climate, it would be foreseeable that the parking lot pavement would become wet, soiled, and contaminated from automotive fluids such as oil, transmission fluid, brake fluid, and so on. Therefore, the defendant had an obligation to provide a safe walking surface under such conditions to their invited guests.

Furthermore, the paint manufacturing industry understands that their products, when used to line a parking lot, may contribute to a slip and fall and therefore recommends the use of an aggregate additive to elevate the slip resistance of their product to be safe under wet or contaminated conditions.

It is my further opinion that the defendants failed in their responsibility to provide a safe walking surface for their invited guests, specifically, Mary, and that each of the defendant's failures served as the direct cause of Mary's slip-and-fall event and subsequent injuries. Had the defendant properly applied the paint to include an appropriate slip-resistant aggregate, Mary would not have slipped and fallen and subsequently injured herself. Furthermore, the defendant had a responsibility to properly maintain (clean) and inspect their premises to ensure the safety of their invited guests, and it appears they did not.

HOW DID IT END?

Mary and the casino settled for a seven-figure amount, and Mary continued going to the casino until it closed a year later.

33

SLIPPERY MUCK!

In early 2014 I was retained in a case involving a woman by the name of Hillary, who in June of 2013 claimed that she slipped and fell on a wet parking lot of a strip shopping center, which was coated with a layer of algae and/or moss. This was right in the middle of a hot Texas summer. How could there be algae growth on the concrete pavement? According to the owner of the strip shopping center, the algae was likely caused by a leaking sprinkling system, which he stated "would leak from the flower beds and collect in puddles in the parking spaces adjacent to the sidewalk."

It is my opinion that multiple failures had occurred that could have prevented Hillary's slip-and-fall event. First, given the proximity of the slip hazard, special care should have been given as to provide a safe walking surface. Many pedestrian slip-and-fall incidents occur on parking lots due to substandard or improperly maintained pavement, which includes cracks, holes, depressions, as well as slippery substances such as automotive fluids, spilled liquids, and organic materials such as moss or algae. Commercial property owners like the defendant have an obligation to protect the general public from unnecessary risk of harm, which includes the proper maintenance of their pedestrian walkways and parking lots.

Secondly, the defendant knew or should have known that their water sprinkling system was broken and was leaking for an extended period into their parking lot directly in front of the walkway in front of the retail store where Hillary slipped and fell. The defendant had both the opportunity and responsibility to mitigate the pedestrian walkway hazard in a timely manner. Based on the photographs provided to me of the walkway in question, it was my opinion that the algae/moss was due to prolonged moisture caused by the leaking water sprinkler system.

The need for safe walkway surface maintenance is clearly outlined in the American Society of Testing and Materials (ASTM) "Standard Practice for Safe Walking Surfaces" ASTM 1637-09, which establishes the standard of care of pedestrian walkways like that of the defendant's parking lot.

Section 5 of the F-1637-09 standard entitled "Walkway Surfaces" outlines the defendant's responsibilities as related to maintaining their exterior walkways. Subsection 5.1.3 requires that "Walkway surfaces shall be slip resistant under expected environmental conditions and use. Painted walkways shall contain an abrasive additive, cross cut grooving, texturing or other appropriate means to render the surface slip resistant where wet conditions may be reasonably foreseeable." Subsection 5.7.1 requires that "Exterior walkways shall be maintained so as to provide safe walking conditions." Subsection 5.7.1.1 requires that "Exterior walkways shall be slip resistant."

The City of Carrollton Code of Ordinances Section 92.50: "Maintenance of private parking and pedestrian areas to be maintained; failure to maintain constitutes nuisance": Section (A) states that "It is a nuisance and shall be deemed unlawful for any owner, lessor or occupant of a premises including, but not limited to, shopping centers, retail establishments, clubs, apartment or office complexes, warehouses, and the like which have vehicle access, parking areas or pedestrian walkways, to maintain such areas or cause such areas to fall into disrepair, either by accident, negligence or purpose, so that the whole or any part thereof becomes a danger to life, limb, or property."

Section (B) states "It shall also be unlawful for any such owner, lessor or occupant to allow the effective use of such areas to become restricted to any degree. Proper maintenance shall provide for the drainage of storm runoff without damage to adjoining property, removal of other liquid wastes and solid debris, removal of dirt deposits and other foreign substances, and removal of tree limbs, brush or other vegetation hanging lower than seven feet above sidewalks or lower than 12 feet above driveways and parking areas. Fire lanes, parking spaces and pedestrian walkways must be clearly delineated. The surfaces of such parking areas and walkways must be preserved in good condition."

Section 92.60 "Causing hazardous conditions or ice to form on streets and alleys": "(A) It shall be unlawful for an owner or occupant to use water or allow or suffer the use of water under their control in a manner that causes the water to collect on or flow across the roadway of a public street, sidewalk or alley and create a hazardous condition, including but not limited to reduced traction, or form ice."

Section 92.61 "Removal of hazardous conditions and ice from sidewalks required": "Every owner, lessee, tenant, occupant, or other person having charge of any building or lot abutting upon any public way or public place shall remove any hazardous condition or ice which has accumulated on the sidewalk in front of or alongside the building or lot as a result of water under the person's control running across the sidewalk and forming a hazardous condition or ice."

It is my opinion that the walkway in question was not in compliance with the ASTM F-1647-09 standard or the City of Carrollton's ordinances named above.

It was my opinion that the wet, algae/moss-coated walkway presented an unreasonable dangerous condition and that the defendant failed to exercise reasonable care in properly maintaining their parking lot and failed in their duty to protect the public, specifically Hillary, from unnecessary risk of injury and were directly responsible for her injuries. Had the defendant (a) properly repaired their in-ground water sprinkling system in a timely fashion as to not allow for the buildup of the slippery algae/moss and (b) power-washed or otherwise removed the accumulated layer of slippery algae/moss, Hillary would not have slipped and fallen and in turn injured herself.

HOW DID IT END?

The case did not go to trial but settled for a six-figure sum. Interestingly, I drove by the strip center about a year later only to find the algae still growing. Sadly, many property owners think that as long as they have insurance, they really don't have to do anything to prevent slips and falls . . . that is, until their insurer drops their coverage!

34

FELL IN THE FREEZER!

For anyone who has ever worked in the food industry, you know that there are potential safety hazards everywhere, including the walk-in freezer. Mitch learned that firsthand one August morning in 2003 when he slipped and fell on ice that had accumulated on a galvanized metal floor of his employer's walk-in freezer.

The freezer floor was wet and/or icy from migrating water located on the floor of the kitchen as well as condensation from the warm-moist air coming from the kitchen. There were no abrasive mats or grates on the floor of the freezer, nor were there any warning signs posted on or near the freezer warning of a wet or icy condition. At the time of his slip and fall Mitch was wearing his company-approved, oil-resistant work boots.

It was my opinion that the restaurant's management exercised poor judgment in their choice not to provide an appropriate floor mat or floor grate in the walk-in freezer to provide a safe walking surface. Smooth-surfaced walkways like that of the galvanized metal floor present a high risk of slip-and-fall injuries when wet or icy and require the application of an appropriate high-traction walking surface. Because no such material was provided, invited guests, including Mitch, were exposed to a significant wet floor slip hazard. The American Society of Testing and Materials (ASTM) has established a walkway safety guideline that requires the use of mats or runners in areas as described above.

Section 5.1.4 of the ASTM F-1637-95-02 "Standard Practice for Safe Walking Surfaces" states that "Interior walkways that are not slip resistant when wet shall be maintained dry during periods of pedestrian use." Section 5.4.2 of the standard states that "Mats should be provided to minimize foreign particles, that may become dangerous to pedestrians particularly on hard smooth floors, from being tracked on floors." Section 5.4.3 states that "Mats or runners should be provided at other wet or contaminated locations, particularly at known transitions from dry locations." Section 4.4.5 states that "Mats, runners and area rugs shall be provided with safe transition from adjacent surfaces and shall be fixed in place or provided with slip resistant backing."

The Occupational Safety and Health Administration (OSHA) CFR 1910.22 requires that "The floor of every workroom shall be maintained in a clean and, as so far as possible, a dry condition. Where wet processes are used, drainage shall be maintained, and false floors, platforms, mats, or other dry standing places should be provided where practicable."

Although Mitch's manager, Kyle, stated that the restaurant used "sandpaper-type" mats in the freezers of some of their other restaurants that were under his supervision, they did not use mats in the freezer at the restaurant in question. The failure to provide such matting was a violation of the industry standards stated above.

A second failure was to warn the incoming pedestrian as to the impending wet floor hazard. There were no warning signs posted adjacent to the doorway at the time of Mitch's visit, nor did the store personnel verbally warn Mitch as to the hazardous walkway conditions that existed within the freezer.

Since the freezer floor was likely to become wet and icy from water migrating from the floor of the kitchen and from condensation, the restaurant's management chose not to provide a remedy to contain such hazards from entering the freezer as well as to provide a slip-resistant walking surface when entering the freezer.

Based on this information, the defendant failed to provide a safe walking and working surface to their employees and therefore failed in their duty to protect their employees from unnecessary risk of a slip and fall. It is clear based on the evidence as well as the experience of the restaurant's management that the floors of the kitchen and freezer are often wet and therefore pose a risk of injury due to a slip and fall. Such knowledge obligates the restaurant's management to safeguard such areas. Unfortunately, they did not, and their failure to do so directly contributed to Mitch's injuries. Had the restaurant's management addressed the above-mentioned wet floor hazard by way of using an appropriate freezer mat or grate along with the posting of a wet floor warning prior to Mitch's entry, it is unlikely that he would have slipped and fallen.

HOW DID IT END?

Mitch's lawsuit was dropped and his medical expenses paid by his employer's state workers' compensation insurance.

35

SLIPPED IN ELVIS'S BATHTUB!

In May of 2016 I was retained to provide opinions in a case involving an elderly woman named May who in the course of showering, slipped and fell in the hotel's bathtub. May's slip and fall occurred back in October of 2013, whereby she injured her right kneecap, left shoulder, and back.

On February 15, 2016, I had conducted an on-site inspection of the bathtub in question.

The hotel was old, built back in the 1950s. The bathtub in room number 303 was also very old and displayed signs of heavy wear. The chrome water drain fixture was labeled Kohler USA, which suggests that the tub was also manufactured by the Kohler Company. The bathtub showed signs that it had once had a textured "slip-resistant" bottom surface, which due to age and excessive wear, had worn smooth, resulting in a dull, gray-colored appearance. Present on the bathtub's surface were four aftermarket plastic "safety strips," which were applied by the defendant to enhance the bathtub's slip resistance. The strips were heavily soiled and smooth to the touch. A rubber bath mat was not provided.

What was interesting to note was the small plaque posted on the outside of room 303 stating "Elvis Slept Here." That's right, the King of Rock and Roll once stayed in the very room and bathed in the very same bathtub that May had slipped and fallen in. The owners said that they have preserved the room in its near original condition, which based on the condition of the bathtub, showed every day of its nearly sixty years of use.

It is my opinion that multiple failures had occurred that could have prevented May's slip-and-fall incident. First, it is my opinion that the bathtub bottom surface was not slip resistant and therefore presented an elevated slip risk to anyone who might bathe/shower in it. This failure was in my opinion the direct cause of May's slip-and-fall event and subsequent injuries.

Such failure to provide a slip-resistant bathing surface is in direct violation of the bathtub industries standard as established by the American Society of Mechanical Engineers (ASME) A112.19.4M-1994 bathtub standard (supplement 1-1998). Although the bathtub was originally constructed to be in compliance with Section 3.3.6 of the standard entitled "Slip Resistance," which requires that "The bathing surface of a bathtub shall be treated in such a manner that it shall comply with the ASTM F462. Treatment shall start 2 in. measured from all side and wall radii and 3 in. measured from the centerline of the drain and from the compound corner radii," the circular slip-resistant discs had worn smooth and no longer provided a slip-resistant bathing surface.

Had the entire bottom surface of the tub-shower enclosure (bathtub) been slip resistant, it is unlikely that May would have slipped and fallen and in turn injured herself. Numerous aftermarket traction-enhancing products such as chemical traction treatments, rubber bath mats, and slip-resistant replacement bathtub liners are available to the hotel industry for the purpose of enhancing the slip resistance of the guestroom bathtubs.

Secondly, during my site inspection I noticed that only a single, vertical grab bar was mounted on the outside of the rear wall of the tub-shower enclosure. It is my opinion that the defendant should have provided at least one vertical grab bar inside of the tub-shower enclosure to assist bathers as they bathe as well as exit/enter the tub-shower enclosure. The installation of grab bars is required by the American Society of Testing and Materials (ASTM) F-446-99 "Standard Consumer Safety Specification for Grab Bars and Accessories Installed in the Bathing Area."

Section 4.2 of the standard requires that "Grab bars are to be installed in the critical support area in accordance with 5.1.1-5.2," whereby the critical support area is defined under Section 2.4 of the standard as "that portion of the back, service, or non-service wall in which support would most likely be beneficial in four different bathing areas (see Figs. 1-4 for specific requirements in this area)."

Section 5.1.1 of the standard requires that "A horizontal grab bar or bars shall be installed on the back-wall critical support area with a total minimum length equal to 30% of the horizontal length of the critical support area." Section 5.1.2 further requires that "A horizontal or vertical grab bar or bars shall be installed in the critical support area on either the service wall or nonservice wall. The horizontal bar or bars shall provide a minimum grippable length of 9 in. (230 mm) (see Fig. 5) within the critical support area. The vertical bar shall provide a minimum grippable length of 6 in. (152 mm) (see Fig. 5) and shall be installed in the tub entrance area."

Based on the above-named grab bar standard, it is clear that the tub-shower enclosure was not in compliance, since there were no grab bars mounted on any of the three critical support areas of the tub-shower enclosure in question.

It was therefore my conclusion that the defendant was not in compliance with the industry standard of safety and failed to provide a reasonably safe bathtub for their invited guests.

HOW DID IT END?

The insurance company for the hotel settled with May. The hotel has since been sold and is being renovated. Elvis's tub has left the building!

36

SHELVED!

On July 29, 2011, Dennis was shopping at a Bay Town, Texas, grocery store, when upon turning the corner of the aisle he was walking in, he suddenly slipped on wire-rack steel shelving that had been removed from the freezer by a store employee. Dennis lost his balance and fell, the results of which led to him fracturing his arm.

Slip, trip, and fall injuries are common in the grocery industry and rank as the leading cause of both guest and employee injuries. Because this fact is well known by the grocery industry, most grocers have formal written procedures outlining their slip, trip, and fall prevention procedures. The defendant had published such safety standards in their *Team Member Handbook*, which stated:

> PDQ Corp is committed to maintaining the safety and health of our team members and customers.
>
> PDQ's management has the responsibility to provide a safe and productive work environment for all team members and a safe and clean shopping environment for our customers. PDQ's management is responsible for monitoring safety compliance and for providing safety assistance whenever needed. Unsafe work habits or conditions jeopardize the health and safety of all team members and customers.
>
> Safe work habits, behaviors, and procedures must be employed when carrying out job duties in all circumstances. All equipment shall be operated in a safe manner according to proper procedure. Any team member who is unsure of how to safely operate assigned equipment, unsure of the company's safety procedures, or unsure of

how to safely carry out any job duty, shall request management assistance before beginning the task.

All team members receive training regarding customer safety, including slip, trip, and fall prevention. Any team member who is unsure of the company procedures for prevention, monitoring, maintenance, and clean-up of any area accessible to customers, including the parking lot, sidewalks, entrances and exits, sales floor, and restrooms shall request management assistance and clarification. Horseplay and careless acts are prohibited because they jeopardize the health and safety of team members and customers. All team members are responsible for reading and understanding this policy and PDQ's commitment to safety. Please clarify any questions or concerns with management. With everyone working as a team, we are confident that we will provide a safe environment for our customers and team members.

ACCIDENT PREVENTION

The handbook also had a section on accident prevention:

Your store was constructed to meet rigid safety standards. Only you, however, can prevent accidents by working safely, watching for hazards and complying with all safety procedures. Report all customer and team member accidents immediately. Report unsafe equipment and other potential causes of accidents also. Studies of team member accidents indicate that the most frequent causes of on-the-job injuries are cuts, falls and injury from lifting improperly.

As we may have thousands of customers shop our stores on any given day, we need to be ever conscious of safety. We believe in a team effort. We ask that each team member use experience, knowledge and good judgment in maintaining a safe and healthful place to work and shop. Housekeeping is extremely important. If you see something on the floor that doesn't belong there, correct it, even if it's not in your department. Team effort is accident prevention at its best. While you work, maintain your "safety attitude" by keeping a few simple work habits in mind.

To which bullet point number five states, "Be observant—you can prevent accidents to fellow team members and customers by watching for and reporting hazards."

INDUSTRY STANDARDS

The National Fire Protection Association (NFPA) Life Safety Code (2003) states in Section 7.1.10.2 entitled "Furnishings and Decorations in Means of Egress," Subsection 7.1.10.2 that "No furnishings, decorations, or other objects shall obstruct exits, access thereto, egress therefrom, or visibility thereof." Subsection 7.1.10.2.1 further states that "No obstruction by railings, barriers, or gates shall divide the means of egress into sections appurtenant to individual rooms, apartments, or other occupied spaces. Where the authority having jurisdiction finds the required path of travel to be obstructed by furniture or other movable objects, the authority shall be permitted to require that such objects be secured out of the way or shall be permitted to require that railings or other permanent barriers be installed to protect the path of travel against encroachment."

The American National Standards Institute (ANSI) A1264.2-2006 "Provision of Slip Resistance on Walking/Working Surfaces" states in Section 8 that "a warning shall be provided whenever a slip or trip hazard has been identified until appropriate corrections can be made, or the area barricaded."

Subsection 8.1.1 Alternate Route: "When there is a slip/fall hazard, which covers an entire walkway, thus making it difficult to safely route personnel around the hazard, barricades shall be used to limit access (see Section 9.1). If appropriate, assign an employee to detour personnel, in conjunction with the appropriate use of warning signs until the barricade can be erected or the hazard removed."

8.4 Placement: "Warning signs shall be placed at approaches to, or around, areas where slip/fall hazards exist. These devices shall surround or be placed around the perimeter of the hazardous area so that it is clear as to where the hazard exists."

Subsection 8.4.1 Unmitigated Hazards: "In cases where hazards cannot be mitigated, warning signs and barricades shall be used to reroute traffic."

E8.4.1: "Warning signs may be ineffective for the control of slip and fall hazards in areas where routing pedestrian traffic around the hazard is difficult to accomplish, such as where the danger lies on a shorter path than the safe route. In such instances, barricading the area in conjunction with appropriate warning signs to reroute employee traffic to an entirely different route may be required. Barricades and warning signs should be used in hazardous areas. See Inherently Slippery Environments Section (Section 9) for details on the use of barricades."

Section 9.1 "Barricades": "Barricades shall be used to isolate processes in hazardous areas. They shall also be used to isolate slip hazards from pedestrian traffic."

9.3 Authorized Entry: "In certain inherently slippery areas, only employees who are properly trained and equipped shall be authorized to enter."

E9.3: "Processes may have to be shut down for special cleanup operations. Equipment may include full personal protective equipment, lifelines, harnesses, or special footwear."

It is my opinion that the defendant, PDQ Corporation, violated their own safety policies as well as those named above and in doing so failed to provide a reasonable standard of care to their invited guests, specifically Dennis. It is clear based on the photographs of the area where Dennis slipped and fell depicting the placement of the wire grate on the floor that a significant and serious slip or trip hazard was posed to individuals coming around the corner of the aisle. It is also clear based on two PDQ store employees' deposition testimony that the PDQ employee did in fact place the wire rack on the aisle way. It is the custom and practice of the retail industry to not intentionally create a hazard by which a customer may slip, trip, and fall, which the employee had clearly done. Having performed maintenance on the freezer unit in question in the past, PDQ's employee was familiar with the maintenance procedure that requires him to remove the wire racks and therefore should have known that it would not be an appropriate practice to place them flat on the aisle way as to expose a customer to a slip, trip, and fall hazard.

Although the store's employee claimed to have posted a wet floor sign near the area he was working, such signage was insufficient to provide a reasonable warning to the impending walkway hazard associated with the metal wire rack. It is the practice to use wet floor signs to warn pedestrians as to a wet floor hazard and not any other type of hazard. Based on his past experience in performing such maintenance work, the store employee should have followed his company's safety procedures as well as those named above by simply barricading the aisle area he was working on as to restrict entry by a customer. Had he done so, it would have been very difficult for customers to enter the work zone as to be exposed to the walkway hazard (i.e., metal rack), which the store employee was well aware of given that he had placed it there.

It was my opinion that Dennis's slip/trip-and-fall event could have been prevented had the store's management properly restricted entry of the aisle in question by way of a barricade. Their failure to do so was, in my opinion, a direct cause of Dennis's fall and subsequent injuries. It is my opinion that the store's management had failed in their duty to exercise good judgment and in turn, exposed their invited guests to unnecessary risk of a slip, trip, and fall.

HOW DID IT END?

The store's insurance company settled with Dennis for $280,000.

37

LOST LEG IN A BATHTUB!

In December 2004 I received a call from an attorney named Marty, who asked if I could help in a case involving a client who slipped and fell in a south Texas motel. "It appears that my client lost his leg because of a slip and fall," Marty said. How could that be? "Well," Marty said, "my client, Bill, was a large man, and while he was taking a shower, he slipped and fell, whereby he landed as to wedge his leg widthwise in the tub. Because of his size and the fact that his leg was broken, he couldn't get out of the tub. Sadly, no one from the hotel entered the room to check on him, so after a few days the circulation in his leg stopped and atrophied."

"After a few days?" I asked. "Why did it take so long for the motel to check on him?" Marty replied, "Well, they didn't have a full-time housekeeper, and the owner's wife, who would fill in, was out of town that week." So for three days Bill, resting in a sitting position with the water controls behind him, remained in the tub. Marty then said that Bill fell after taking his shower and that the water was off. He had nothing to eat and was severely dehydrated.

I was retained the following day and began reviewing that case. Marty scheduled a site inspection, and off we went to the Lazy-J Motel. Once in the room where Bill fell, I couldn't help but notice a foul stench. The combination of cigarette smoke, mold, mildew, and whatever the guest the night before ate lingered. The smell was so bad that a day later my clothes stilled reeked of room number 102 of the Lazy-J.

Upon inspecting the bathroom, I noticed that the tub in question was a bit smaller than expected. "That's how they made them back in the day," the owner said. Also noticeable was the fact that there was no slip-resistant pattern on the bottom of the tub, which is a violation of the industry standard as published by the American Society of Mechanical Engineers (ASME), ASME standard number A112.19.4M-1994 (supplement 1-1998), which in Section 3.3.6 "Slip Resistance" requires that "The bathing surface of a bathtub shall be treated in such a manner that it shall comply with the ASTM F462. Treatment shall start 2 in. measured from all side and wall radii and 3 in. measured from the centerline of the drain and from the compound corner radii."

I had also noticed that there were no grab bars on the wall. Although grab bars are not mandatory in all hotel rooms other than those designated as handicapped, they are an inexpensive means by which to assist anyone who may experience a slippery bathing surface. A growing number of hotel owners across the country have enacted this policy, which the Hotel Association's Loss Prevention and Executive Engineers and Environmental Committees have openly supported.

During my inspection, I had noticed what appeared to be a brand-new white rubber bath mat draped over the edge of the bathtub. Mats of this type are customarily used by the hotel industry and serve to enhance the slip resistance of their bathtubs. It is my understanding based on Bill's deposition testimony that no such mat had been provided upon his stay. Had such a mat been provided by the motel's management, Bill's slip-and-fall incident may very well have not occurred.

In his deposition testimony Bill stated that the tub surface was "very slippery" and that "there was no bar or anything to hold onto." This problem is common in the hotel industry, which experiences hundreds of slip-and-fall accidents each year. In fact, bathtub slips and falls are the leading cause of guest accidents for the hospitality industry. In his deposition testimony, the owner of the motel acknowledged this fact by stating that "bathtubs are going to be slippery" and that they can be made safer by using rubber bath mats, strips, or grab bars. Although the owner of the motel acknowledged both the problem and the solution, he chose not to provide any such preventative measures.

During my inspection, I had the opportunity to question the housekeeper, Elise, who was off the week Bill slipped and fell. I asked her how she went about cleaning the motel room bathtubs. She responded by saying that all the supplies she was provided to clean each room were contained in her cleaning tray, which had three containers. She then went on to describe the three-step procedure by which she applies a chemical called KABAM to the bathtub. She then brushed the surface with the white-handled toilet brush and then dried the tub with a rag. Needless to say, this cleaning procedure is not customary to the industry and does not comply with the industry standard of care for bathing surfaces. Who cleans bathtubs with the same brush used to clean the toilet?

The motel owner acknowledged that he did not have any procedures or policies for maintaining and cleaning the bathtubs and that Elise was not given any training, just a tray of cleaning products, rags, and a brush.

The cleaning process described by Elise may have served to make the bathtub more slippery and certainly less sanitary. The cleaning chemical used by Elise was not a task-specific product (e.g., tub and tile cleaner) but rather a generic "all-purpose" product. Products like KABAM often leave a slippery film that can build up over time, making surfaces slippery, especially when wet. The KABAM product is not an appropriate cleaning chemical for bathtubs and should not have been given to Elise to clean bathtubs. Furthermore, Elise may be inadvertently contaminating each of the bathtubs by using a toilet bowl brush rather than the soft nylon utility brush she had in her room tray. Cross-contamination may also be spread by using the dirty rag she claimed to have used to dry the bathtub. Such failure to train their housekeeping staff had compromised the safety and health of the guests.

Based on this information it was clear that the management of the Lazy-J Motel had failed in their responsibility to protect their invited guests from unnecessary risk.

It was also my opinion that Bill's slip-and-fall incident could have been prevented had the tub been provided with grab rails and properly safety treated and maintained in a manner as prescribed by the hotel industry. Had the motel's management addressed the bathtub hazard prior to Bill's visit, he would not have slipped and fallen.

HOW DID IT END?

A few months after I presented my report, the case settled for $850,000. Bill indeed lost his leg because of a slip and fall. Although this outcome is rare, it is sadly very common for individuals, especially the elderly, to die as a result of a bathtub slip and fall.

38

AS SEEN ON TV!

In August of 2014 I was retained by a homeowner by the name of Marge, a middle-aged woman who lived with her husband, Bill, in a quiet town in Georgia. Marge, like thousands of Americans, saw the television ad for the Aqua Rug and thought it would be a good idea to buy one given that her bathtub was many years old and had become a bit slippery. Marge, like many women, shaves her legs while showering, which means that she has to stand on one leg at a time, thus increasing the chances of a slip and fall. Marge purchased the Aqua Rug at a local retail store and began using it the following day. The first thing she noticed was how comfortable the woven surface was underfoot, but the bottom-mounted suction cups did not grab her bathtub's surface very well. A few weeks passed, and the problem with the Aqua Rug's suction cups seemed to be getting worse. The mat began getting stiff and was now sliding on her tub even more, until one day while she was showering, it slipped, causing Marge to fall in her bathtub and injure her shoulder and arm.

In the summer of 2014 I was retained by Marge to provide an expert opinion as to the safety of the Aqua Rug product and its appropriateness as a bath mat. Later that summer I began to do some research on the Aqua Rug. My research revealed that the Aqua Rug product in question was manufactured in Taizhou City in Zhejiang Province, China, and distributed by Alibaba.com, which identifies the product as a "hot selling shower rug/pvc bath rug/water rug." The identical mat is distributed by idealpanda.en.alibaba.com as the "Aqua Rug," which describes the material as "100% Acrylic" (plastic) with an "Anti-Slip feature."

The Aqua Rug material consists of two layers, a top-spun PVC plastic or acrylic material and a heat-fused, clear, checkerboard-patterned PVC backing. The Aqua Rug mat measures approximately 29 3/8 inches (75 cm) in length by 16 5/8 inches (43 cm) in width and 5/16 inch thick (10 cm).

The manufacturer did not provide any test data as it related to the mat's slip resistance. This was particularly interesting given that the mat was patented. According to the section entitled "Field of the Invention" in the U.S. Patent number US2012/0094057 A1 dated April 19, 2012, which was issued to Mr. Joel Patrick Bartlett for his "Porous Anti-Slip Floor Covering," one such trademark use is that of the "Aqua Rug" bath mat. The material was patented as a "floor covering material and particularly suited for moisture-prone applications and that provides superior traction and comfort" (see appendix). To quantify the slip-resistant characteristics of the Aqua Rug's synthetic backing and suction-cupped edge, two 3-inch-by-3-inch sections of the virgin exemplar Aqua Rugs were cut out and tested per the National Floor Safety Institute's (NFSI) 101-C Test Method for Measuring Dry TCOF of Floor Mat Backing Materials. Testing was performed under wet (distilled water), wet (SLS solution), and dry conditions, the results of which are listed below:

Based on the slip-resistance test results of the exemplar Aqua Rug, it was my opinion that the product's backing material when new offers a sufficient level of transitional (stepping) slip resistance under dry conditions but does not under wet or soapy/wet (SLS) conditions like that which occur during the bathing/showering process.

The manufacturer's instructions specifically call for applying the Aqua Rug directly, to fill the bathtub or shower with a small amount of water, place the Aqua Rug in the tub, and press down suction cups to secure. This procedure is designed to affix the Aqua Rug firmly to the surface of the bathtub or shower. As a part of my investigation, I attempted to replicate the use of the exemplar Aqua Rug on three different bathtub/shower units. What I experienced was that when I used the Aqua Rug according to the manufacturer's instructions, it did not move upon application to the wet bathtub/shower; however, shortly after stepping onto the Aqua Rug it began to buckle and eventually came loose. The sliding sensation underfoot felt like being on a surfboard.

Although the suction cups remained adhered to the bathtub/shower's surface, they began to slide laterally along the tub's surface and eventually released. Slippage was further amplified as I scrubbed my feet on the surface, which is one of the selling features of the Aqua Rug.

It is my opinion that the Aqua Rug product has several design flaws, the first of which is the size, number, and location of the four 1 3/8-inch round vinyl suction cups located at each corner of the Aqua Rug. Under normal use, the suction cups simply do not provide for enough adhesive force to keep the mat from moving. Secondly, the PVC backing material is not securely affixed to the spun woven surface and when stepped on will induce slippage between the mat and the bathing surface, further compounding the mat's overall movement. It is my opinion that these two design flaws served as the proximate cause of Marge's slip-and-fall event.

The product nor its packaging as provided by Tristar Products, Inc., did not contain any cautions or warnings as to the slippery nature of the product nor did it provide any tips for keeping the product adhered to the tub surface, such as the need for a clean bath surface before application. Furthermore, the manufacturer's instructions failed to provide any specific cleaning procedures for the mat's backing and suction cups as to enhance its slip resistance. This is particularly important given that upon normal use, numerous bath oils, soaps, and creams are used that may compromise the mat's ability to remain stable and affixed to the bathing surface.

Tristar Products' marketing materials state that "the Aqua Rug is the first quality carpet uniquely made for your shower, tub or anywhere there might be water, mildew, dampness or dirt" and that it "is a pleasure to stand on and helps you feel secure in slippery spots. Its no-slip backing guarantees it will stay put wherever you place it." Tristar Products further stated that "unlike traditional bath mats and no-slip rubber bath inserts, Aqua Rug will never stain, form mildew or develop bacteria."

In a side-by-side visual comparison of the actual mat purchased and used by Marge to that of an identical new mat I purchased, it is clear that these claims are unsupported. The subject mat had clearly changed color and was stained as to be a beige color. Furthermore, microscopic photographic images reveal that the subject mat's patterned backing had separated from the spun woven surface and displayed evidence of mildew growth.

The Aqua Rug product is marketed and sold from Tristar Products, Inc., which is based in New Jersey. It is my opinion that Tristar Products, Inc., misrepresented and falsely promoted the slip-resistant characteristics of their Aqua Rug product and should have provided a warning label on the actual product informing their customers as to the potential danger they would be exposed to when utilizing the Aqua Rug product as prescribed. Although the product literature did contain a caution message stating "Do not use on textured or uneven surfaces, may cause slip injury," such message was not affixed to the actual mat and therefore was not an appropriate warning. Such failure to properly warn the consumer of a safety hazard is a violation of the Consumer Products Safety Commission's (CPSC) Manufacturers Guide to Developing Consumer Product Instruction (October 2003) and the American National Standards Institute's (ANSI) Z535.6 Standard for Product Safety in Product Manuals, Instructions, and Other Collateral Materials.

It was my conclusion that the Aqua Rug product did not provide an adequate level of traction as advertised and when used per the manufacturer's instructions could present a slip hazard.

HOW DID IT END?

Shortly after producing my report the case settled for an undisclosed sum. Interestingly I was retained in another Aqua Rug matter. The retailer who sold the Aqua Rug to Marge no longer carries the item. The promoter, Tristar, settled with Marge for an undisclosed sum.

In January 28, 2016, the U.S. Consumer Product Safety Commission (CPSC) had issued a recall of the Aqua Rug product (see appendix), stating that "The four suction cups on the underside of the rugs can fail to prevent slipping, posing a fall hazard to the user." According to the CPSC they claimed to have received sixty reports from consumers who had fallen in their tub or shower due to the Aqua Rug's suction cup failure. Of the sixty cases, thirty people reported injuries that were similar to Marge's.

Consumers were instructed by the CPSC to "immediately stop using the recalled shower rugs and contact Tristar for instructions on how to dispose of the rugs and to obtain a free replacement rug."

39

TANKED!

On a cold December night in 2003 Darrell was working as a night watchman at a food manufacturing plant in Dallas, Texas. His job called for him to perform an inspection called "walk the yard," whereby Darrell walked the entire exterior perimeter of the plant once an hour. Flashlight in hand, Darrell went about walking the yard. He had just started his shift and noticed that the company was having some new construction to one of their outside storage tanks. As Darrell walked along the construction site, he suddenly fell into a hole beneath a dark-colored tarp. This just wasn't any hole but was a six-foot-deep underground chemical waste tank that was opened as part of the new construction. Rather than covering the tank with a steel plate, the construction company simply draped a plastic tarp across it. Having broken his legs, Darrell remained in the tank for the remainder of the night. It wasn't until the crew arrived the following morning that they discovered Darrell nearly frozen to death.

The waste tank was not barricaded nor were warning signs posted alerting pedestrians as to the impending fall hazard. Because the location of the incident was a construction zone, the food company and their contractor should have provided proper warnings as to both the potential and known hazards (i.e., walkway hazards) the construction zone presented. This failure to warn is a direct violation of the Occupational Safety and Health Administration's (OSHA) Code of Federal Regulations (CFR) 29, Section 1910.145 entitled "Specifications for accident prevention signs and tags."

Secondly, the property owner and their contractor(s) were aware that the tank hole was present in the walkway but did not properly barricade it, nor did they provide protective guardrails or an appropriate cover. These failures are in violation of the General Requirement section of OSHA CFR 29, Section 1910.22 and the Guarding Floor and Wall Openings and Holes Section 1910.23. Furthermore, these failures to provide a safe walking surface are in direct violation of the National Fire Protection Association (NFPA) Life Safety Code. Section 7.1.8 of the Life Safety Code requires that guards "shall be provided at the open sides of means of egress that exceed 760 mm (30 inches) above the floor or grade below." This requirement to guard deep walkway openings is echoed in Section 11.1.8 of the NFPA 5000 Building Construction and Safety Code.

Finally, the failure to provide a safe walking surface was a direct violation of the American Society of Testing and Materials (ASTM) standard F-1637-02. Section 5.1.2 states that "Walkway surfaces for pedestrians shall be capable of sustaining intended loads." The walkway in question was clearly in violation of this standard. All of the codes listed above were in place at the time of Darrell's fall.

It is my conclusion that the plastic tarp that was draped over the waste tank hole served only to camouflage the impending danger of a person accidentally falling into the hole. By doing such, the property owner and their contractor created an unreasonably dangerous and unsafe condition. Had the property owner and/or its contractor properly barricaded, guarded, and warned of the known walkway hazard prior to Darrell's visit, it is most likely that he would not have fallen into the hole, which was the source of his injuries. It is my opinion that both the defendant and their contractor were negligent in their duty to protect the public from unnecessary injury and were directly responsible for Darrell's injuries.

HOW DID IT END?

The case did not go to trial but was settled for an undisclosed six-figure sum.

40

CHICKEN RUN!

One of the first lawsuits I was retained in dates back to 1997 and involved a young girl who slipped and fell in an East Texas chicken restaurant. Her attorney called me, and at the time I was employed as the president of Traction Plus, Inc., which was a company I started back in 1990, specializing in producing slip-and-fall-prevention products. Having never provided opinions related to slip-and-fall lawsuits, I told the attorney that I was probably not the right person for the job; he responded, "You're an expert in slip-and-fall prevention, and that's exactly the kind of person I need." What I didn't realize was that this "out of the blue" telephone call would change my life and serve as the foundation of an emerging new career as a forensic expert.

The attorney described the case as follows: "I represent a young girl by the name of Annie who slipped and fell in the kitchen of a chicken restaurant and got badly burned." Well, having some experience selling floor safety products to the restaurant industry, I felt a bit more comfortable taking the case. The attorney went on to tell me the story that one night after the restaurant closed, Annie and her fellow workers were cleaning up the restaurant when the manager, who was mopping the kitchen floor, said, "Hey, everyone, let's have a contest to see who can slide on the wet kitchen floor the farthest." Everyone that has ever worked in a restaurant knows just how slippery their kitchen floors can get. As innocent as it sounded, Annie joined the game. After a few of her coworkers took their shot at sliding on the floor, it was Annie's turn. She backed all the way to the rear door of the kitchen, moving empty boxes to ensure nothing was in her way, and off she went. As she began to slide three, now five, then six feet, she suddenly hit a dry patch of the quarry-tiled floor, which brought her to an abrupt stop. The sudden stop caused her to propel herself forward, whereby her outreached arm landed in a four-hundred-plus-degree pressure-cooker deep-fryer. Her hand and arm up to her elbow submerged completely into the deep-fryer. As Annie pulled her arm out of the deep-fryer, the palm of her hand had been torn off by the extreme heat of the fryer's heating element.

At the time, I had a daughter also named Annie who was seven years old. This young girl was someone's daughter, my daughter and your daughter. This case now became personal to me and changed my life forever. Once retained, I wanted to make sure that I didn't screw up this girl's life any more than it had already been hurt. I had to get it right. I have to do my best, and that means being thorough and fair to all parties involved. I worked with restaurant workers every week and sold Traction Plus products into a growing number of restaurant chains. I knew that slips and falls were the leading cause of injury in the restaurant industry, but this case was different. It wasn't just an injury statistic that I often quoted in my sales pitch; it was a real person.

I always questioned why it is that we send our children off to work in businesses that expose them to countless safety hazards like sharp knives, meat slicers, and automatic equipment. It's common for people to burn themselves in kitchens, but this was different. This was really bad!

Once I received the images of Annie's burn, I cringed. This sweet little East Texas sixteen-year-old was forever permanently injured, and the image of her burn was burned into my memory for life. The doctors said that the skin grafts should work, and with physical therapy she should regain partial use of her arm and hand, but the emotional trauma will last her a lifetime. Her position on her high school cheerleading squad was over. Although she is still a member, she cannot participate. Her self-image is forever changed. Changed due to a "who can slide the farthest contest" at the local chicken restaurant.

THE LAWSUIT

I was retained to provide opinions as to the appropriateness of the restaurant's kitchen floor, floor maintenance products and procedures, and the overall safety of the operation. The kitchen floor was of six-inch-by-six-inch red quarry tile, the type of tile I had seen (and cleaned) hundreds of times in restaurant kitchens. The restaurant used a common degreaser to clean their floors, which they purchased from their local wholesale club store. They cleaned all their floors via the traditional mop and bucket approach. At first glance it looked like their floor and cleaning process was typical for a restaurant. But I knew that the typical routine of cleaning restaurant floors was not the right way. In fact, improper maintenance was a leading cause of kitchen slip-and-fall events.

It all starts with the floor, red quarry tile. Although common in restaurant kitchens, it's one of the worst choices. Most restaurants use quarry tile because it's very durable and very cheap. Quarry tile is made from clay and has a very high absorption rate, meaning that whatever gets spilled on it, like water, oil, and food, a portion will be absorbed into the pores of the tile, making it difficult to remove. Over time a wax-like coating builds up and is very hard to remove. These polymerized films often create a shiny appearance, which is desired by the owner, but polymerized films are extremely slippery when wet and even more so when exposed to soap and water like that used by the restaurant. To make the situation even worse, the restaurant, like many others, mopped the floor using a cotton mop head. Cotton mops are also very absorbent and will hold within their fibers the very contaminants they are intended to remove. Mopping a quarry-tiled floor with soap-based degreaser and a dirty mop head is akin to painting the floor with the crud you just mopped up from an adjacent section of flooring. You cannot properly clean a quarry-tiled kitchen floor using a conventional degreaser and cotton mop head. I knew this from my experience with Traction Plus.

Secondly, neither Annie nor any of her coworkers were wearing slip-resistant shoes. Today slip-resistant footwear is an expectation in the restaurant industry, but back in 1997 it wasn't. My experience in the slip-resistant footwear industry again came about through my experience with Walmart Stores, who sold a brand of slip-resistant shoes called Traction Plus/TRED SAFE. The primary customers for the product were restaurants, and for under $20 a pair, it was an affordable way for folks who work in restaurants to not slip. Had Annie or her coworkers worn slip-resistant shoes, it's unlikely that (a) the manager would have had a slip-on-the-floor contest and (b) that Annie's tragic slip-and-fall event would have occurred.

It was my opinion that the cause of Annie's slip and fall was based on a combination of failures. First, it was a bad idea to host such a foolish contest and spoke to the failure of her manager to understand and accept responsibility for his workers' safety. Secondly, the restaurant kitchen floor was one that was difficult to clean, and the method used to mop the floor was inadequate. The restaurant should have cleaned the floor by flooding it with a solution of cleaner and hot water and allowed the solution time to work. Once the floor was flooded with cleaning solution, deck brushing should commence. The agitation action of deck brushing is the best way to remove polymerized films. Using a traction-enhancing cleaner also helps. Lastly, they should have rinsed the floor with clean water and squeegeed all remaining water and solution into the floor drain. The problem with this type of procedure is that it's hard work and very time consuming. My experience in the restaurant industry told me that most workers take shortcuts when it comes to cleaning the floors, and the end result is a slippery floor.

HOW DID IT END?

The case did not go to trial but settled for a high six-figure sum. The restaurant's insurance company was smart to know that they didn't want to take the case to trial and expose Annie's horrific and disfiguring injury to a group of East Texas jurors. Regardless of the settlement she received, it was clear that no sum of money would ever make Annie or her family whole again. Annie's slip and fall changed her life forever and mine as well.

41

A BAD SIGN!

It was January of 2005 when I wrote my report for my first international case. Having been recently retained by a defense firm out of Toronto, Canada, I was interested in learning how the tort system of law worked in Canada.

The claim being made by the plaintiff, Kelsey, was that on the day after Christmas 2000 in Toronto, when Kelsey and her son were shopping at their local big box retailer, she turned the corner of an aisle and tripped over a short, yellow wet-floor sign that was just behind a floor display, which she claimed obstructed her view of the sign.

The alleged trip and fall occurred on the vinyl tiled walkway located near the candle aisle of the big box retailer in question. The walkway surface was dry and in good condition and clear of any foreign objects. The caution sign was not broken, torn, soiled, or otherwise defective. The sign that Kelsey claimed to have fallen over was a short, orange-colored, pop-open type. The sign was lightweight and designed like a pop-open tent. It was hard to tell how this type of sign could have caused her to trip and fall.

THE LAWSUIT

Kelsey was suing both the big box retailer and the manufacturer of the sign, the ACME Company, claiming that the placement and design of the wet-floor sign presented an inherent danger to pedestrians. I was retained by both codefendants.

Kelsey's attorney hired an expert by the name of Walter, who referenced the "Ontario Retail Accident Prevention Association (ORAPA) guidelines" for increasing customer safety. Having some familiarity with the ORAPA, I knew that although informative, ORAPA is not a recognized authority in pedestrian safety nor do they claim to offer standards of such; rather, as Walter stated, they submit non-industry-specific "guidelines" or recommendations. These guidelines are often described in very general terms and are not legally enforceable.

Furthermore, it is my view that based on all the evidence presented to me at that time that the display was not related in any way to Kelsey's trip-and-fall incident. Neither Kelsey nor any of the deponents in this case made reference to the display's being a contributing factor let alone a direct cause of her incident; therefore, Walter's suggestion that the display was "an underlying cause" is unfounded and without merit.

Walter also had suggested that the "caution wet floor" marker presented a "significant tripping hazard to customers" and referenced a National Safety Council (NSC) statement that the sign be twice as tall. As a member of the National Safety Council, I am not aware of any such statement being published and submit that any such opinion is not enforceable, since the NSC is not a standards-creating organization nor do they propose to be an authoritative body on the manufacture and/or use of wet-floor signs. The statement that Walter had referenced appears to be an editorial recommendation and is not in any way to be applied as a standard. In fact, to my knowledge, there are no specific standards related to the design, construction, or size of wet-floor signs.

Walter's suggestion that an alternate-size sign be used is again not based on anything other than his opinion. Caution signs are manufactured in a wide range of colors, styles, and sizes. It is the property owner's responsibility to specify what they consider to be an appropriate sign for the given application that is used. I use as a basis of this opinion the words of Kelsey herself, who stated that "there were lots of people on both sides" (of the aisle), all of whom were able to safely navigate around the exact caution sign that Kelsey claims was a tripping hazard. If in fact the sign presented a "significant tripping hazard to customers," then one would expect others to have experienced a similar trip-and-fall event, and no such event has taken place. In fact, according to the manufacturer of the sign, no claim for a trip-and-fall incident has ever been reported in the company's history.

It was my opinion that the sign in question was well designed and safe for use. To illustrate this point, I submitted a similar type and size floor sign as that in question, which was manufactured by a different company and was identical to that of the product manufactured by ACME.

It is my conclusion that Kelsey's fall was most likely the result of her preoccupation with shopping. Both Kelsey and her son stated that the store was crowded due to the post-Christmas (Boxing Day) sales event. Each of the deponents had described heavy traffic in the aisle in question, which forced shoppers to cluster together and migrate slowly. The deponents also described other shoppers, including Kelsey's son, migrating safely around the sign in question. It is my opinion that Kelsey simply did not see the caution sign, which was placed directly in front of her as she made her way down the aisle. Such is common when pedestrians are preoccupied with other tasks.

Caution signs are used specifically for the purpose of drawing a pedestrian's attention to a pending hazard, in this case, a potentially wet floor hazard. Signs of the type used are commonly used in the retail industry and do not pose a hazard to pedestrians.

It was therefore my conclusion that neither the manufacturer of the caution sign nor the big box retailer were negligent in their duties to provide a safe walking aisle way to their invited guests, and although unfortunate, Kelsey's fall was not the result of a defective product or misplacement.

HOW DID IT END?

The court in Canada discharged the case on summary judgment, saying that there was not enough evidence for the court to move forward.

42

RAILROADED!

Jack was a railroad mailman who loved traveling the country via the rail. It was in his blood. Jack came from a long line of railroad people and was familiar with all the western states' rail lines. January 30, 2001, was just an ordinary day for Jack. He boarded rail car #1725 and began his duties sorting mailbags. As Jack was walking across the car, he suddenly found himself on the ground. It was a puddle of water located on the platform of the mail car that caused his slip and fall, whereby he landed on his back near the top nosing of the stairway. The injury was severe: a dislocated disk and a broken vertebra. Because Jack had a previous back injury, his fall was that much more severe.

The flooring material was a heavily worn section of stainless steel diamond plate, which was installed in approximately 1960. There were no abrasive/slip-resistant stair nosings or nonskid safety strips installed on the steps or platform floor. The platform floor was wet from the accumulation of rainwater from a previous storm, which the previous workers did not clean up. Because of poor overhead lighting, Jack did not know the platform floor was wet prior to his fall. Warning signs were not posted on or near the walkway hazard.

The Occupational Safety and Health Administration (OSHA) Code of Federal Regulations (CFR) 1910.22 "General Requirements," which applies to all permanent places of employment, except where domestic, mining, or agricultural work only is performed, requires that:

1. All places of employment, passageways, storerooms, and service rooms shall be kept clean and orderly and in a sanitary condition.
2. The floor of every workroom shall be maintained in a clean and as so far as possible, a dry condition. When wet processes are used, drainage shall be maintained, and false floors, platforms, mats, or other dry standing places should be provided where practicable.

The Occupational Health and Safety Administration (OSHA) CFR 1910.24, Section (f), requires that "All stair treads be reasonably slip resistant and the nosings shall be of nonslip finish."

The American Society of Testing and Materials (ASTM) F-1637-95, Section 6.1.2 "Standard Practice for Safe Walking Surfaces" requires that "Step nosings shall be readily discernable, slip resistant, and adequately demarcated."

The American National Standards Institute (ANSI) A1264.1-1995, Section 6.7 entitled "Slip Resistance" requires that "All treads and nosings shall be of slip resistant material."

Section 49 CFR Chapter II (10-1-99 Edition), Section 238.305 (C) (2) entitled "Interior calendar day mechanical inspection of passenger cars" and Section 238.307 (C) entitled "Periodic mechanical inspection of passenger cars and unpowered vehicles used in passenger trains" requires "As part of the periodic mechanical inspection the railroad shall verify the condition of the following interior and exterior mechanical components, which shall be inspected not less frequently than every 92 days. At a minimum, this inspection shall determine that: (1) Floors of passageways and compartments are free from oil, water, waste, or any obstruction that creates a slipping, tripping, or fire hazard, and floors are properly treated to provide secure footing."

This was a unique case where dozens of workers were deposed, which served as a foundation for Jack's attorney to build their case. The broad contempt for safety that Jack's coworkers and supervisors had was amazing.

DEPOSITION TESTIMONY

Kevin was the head of maintenance for the railroad and stated that in his opinion, the train car floors were not in compliance with Section

238.307 (c) (1) of the code of federal regulations and were not properly treated as to provide secure footing. He said that he was not trained in floor safety, but it was his understanding that when the floors were wet, the diamond plate material would provide an increased amount of slip resistance. He was unaware of any written procedures for inspections of train cars back in 2001 and was of the understanding that nonskid strips were applied every three years as a part of the railroad's replacement program. He said that he had no way of knowing if the steps and platforms were worn and therefore hazardous and that the train cars had to wait until they were up for replacement before they were to have safety strips applied. A floor with a slip hazard would not be in compliance with Section 238.305 (c)(2). He confirmed that train car #1725 received new nonskid strips on August 4, 2003.

Tom was Jack's supervisor and stated that new sanded strip steps were a good idea and were included as a recommendation in his reports and that he reported worn strip conditions prior to 2001. He also said that 50 percent of the train cars had strips and that the diamond plate steps and platform in car #1725 were worn down; he had personally slipped on the diamond plate surface(s). Tom admitted that it is difficult to see water on the diamond plate material and that the lighting on the train cars was poor.

Harry was a coworker and union representative for the railroad, recalled attending safety captains' meetings, and stated that the union had requested that safety strips be applied to the step nosings before 2001, but they were not installed.

Rick was a company union representative who stated that no train cars had "nonskid" applied back in 2001 and that the decision to add strips was made in 2000. He also stated that he was unaware of slip-and-fall complaints from safety captains' meetings.

Joe was the train's engineer at the time of Jack's slip-and-fall event and recalled that the floors in the train cars were slippery when wet and that he had observed train passengers slipping.

Greg was the company's safety manager and stated that he had completed the Team Accident Investigation form but did not get involved in the actual investigation into Jack's slip-and-fall event. Greg further stated that Jack had violated rule number 80.1 and 80.2 of the company's safety handbook, which calls for employees to avoid creating hazards and clean up after themselves, a requirement Jack was unaware of and a hazard that Jack had not created. Greg also said that he did not speak to Jack prior to authoring the TAI, which is in and of itself a violation of the company's accident-reporting procedure.

Justin was a maintenance worker or "cleaner" for the railroad and stated that the job of the cleaner is to clean the train cars and mop up any water or spills, and the car man then inspects. Rule 80.1 requires that accumulated water in the vestibules is to be mopped up, except for the mail car, which is not cleaned. If a train car was missing a nonskid strip, the car man would note it on the 238 inspection form.

Dan was the railroad's risk manager, and he stated that slips, trips, and falls were a major and ongoing issue and that the company was targeting this issue as a main source of injuries; slips and falls had been an issue for a year or two prior to 2001. Prior to 1999, strips were applied to the trains. He was aware that the diamond plate material became slippery over time, and in his view the diamond plate surfaces were "very slippery," which would cause people to "move on it against your will" and that he had personally slipped down the diamond plated stairs and seen others do the same.

He went on to say that the union recommended the application of strips as a minimum step toward safety but that the railroad company refused to speed up the process of applying the strips and relied upon the three-year railroad car repair cycle; the reason for delaying the application of strips is due to the fact that the company did not want to invest money in a car that was slated to be taken off line soon. Dan said that the vestibule area where Jack slipped and fell was considered a slipping "red zone" and that the company did not always have the same people attend the safety captains' meetings, which "was a great source of frustration." Furthermore, Dan said that the company did not communicate with the union as to the products they were considering. A quarter to a third of the train cars did not have strips installed. Both the union and the railroad company had copies of the safety meeting minutes and were aware of the need for strips. According to the committee's June 21, 2000, meeting minutes, the problem of "slipping on trains' diamond plate" was specifically addressed.

Another mailman by the name of Craig stated that in 2001, if he had observed that the diamond plate material was excessively worn, he would do nothing to correct it. He would have done so for missing slip strips. He also said that the railroad company's repair policy went from a two-year cycle to a four-year cycle back in 2001. Strips were being applied to stairs prior to January 30, 2001.

Dave was a safety captain for the railroad and said that the subject of the anti-skid coating was first brought up by the safety captains back in 1999. He was aware of people slipping on steps and had witnessed mist coming through the door openings. A person by the name of Cushman from the transportation department was responsible for the application of the nonskid strips and was unaware of the details of the products being tested.

Hank was Jack's immediate supervisor who interviewed him after his slip and fall and asked him to complete an incident report. Hank did not inspect the train car nor did he interview anybody regarding Jack's incident, stating that in 2001 there was no requirement to report vestibules that did not have nonskid strips, and therefore the absence of nonskid strips would not be a cause for an exception. Hank agreed that the purpose of train car inspections was to ensure that the car was safe for use but that he did not know if car #1725 had strips in it or not at the time Jack slipped and fell. He did not personally inspect the car, and therefore he did not recognize the absence of nonskid strips as a potential hazard. Hank concluded by saying that he was unaware of any complaints or discussion about the absence of nonskid strips prior to the date of Jack's slip and fall.

It is my opinion that like most slip-and-fall events, multiple failures had occurred that could have prevented Jack's slip-and-fall event. It is my view that the fact that the platform as well as the adjacent steps were wet increased the risk of a slip-and-fall incident. This opinion is echoed by generally accepted safety industry literature, including that published by the American Society of Safety Engineers (ASSE), which states, "Stairs that are wet or cluttered may be more hazardous. Sand, mud, algae, ice, snow, water, poorly maintained carpet or foreign objects may cause people to lose their balance if they step on them. The problem may be magnified in cases where the lighting is poor and the objects cannot be clearly seen or anticipated." The train car cleaner and car man apparently failed in their duty to remove the accumulated water from the surface of the platform of train #1725 prior to Jack's entry. Justin's testimony was that because car #1725 was a mail car, it was not to be maintained like a passenger car.

A second opinion was that although metal diamond plate is commonly used in train cars like that named above, such material, over time, will wear down as to produce a slippery walking surface. Worn diamond plate, like that used in car #1725, does not provide an adequate level of slip resistance and in turn presents an unreasonable risk of harm to anyone walking on such a surface. Such performance deficiency is further compounded by the application of water, ice, or snow. It is the obligation of the railroad company to ensure that the stairs and walkways within their train cars are maintained in a safe condition. Although diamond plate gives the appearance of providing a high degree of traction, it typically does not. Although there are certain kinds of diamond plate materials that do actually provide adequate traction even when wet (e.g., razor edge), the diamond plate platform and stairs in the above-named case was not of that type. Furthermore, both the company and union were aware of the dangerous conditions being experienced on the diamond plate as reflected in the June 21, 2000, meeting minutes. A third and perhaps most important opinion addressed was the absence of appropriate hazard identification. As the front-line defense in the prevention of slip-and-fall accidents, hazard identification is an important tool. It is my understanding that wet-floor signs were not posted at or near the location of the wet-floor hazard to which Jack was exposed. It is the obligation of the railroad company and their management to warn both their employees and their invited guests as to known hazards or potential hazards, which the train company failed to do.

Lastly, given that the railroad was aware that their employees would be exposed to wet walking surfaces as a part of their job, the company had a responsibility to provide appropriate "slip-resistant" footwear to their workers. Having read the relevant sections of the railroad's safety guide that described their footwear policy, I find no recommendation or requirement that employees of the railroad wear slip-resistant footwear. In fact, there is no mention of the term "slip-resistant" anywhere within the company's footwear policy. The company's failure to provide adequate information as it relates to proper (i.e., slip-resistant) footwear placed their employees at risk of a slip-and-fall event.

The incident report prepared by Greg stated the "primary cause" of Jack's slip-and-fall incident was that the "employee failed to be alert of footing conditions." The report then stated that the "contributing factor" was that of a "wet vestibule platform." It is my opinion that Greg had completed the incident report incorrectly and misidentified the true cause and contributing factors. The cause of Jack's slip and fall was not that he did not see the wet-floor hazard but rather that the platform floor was wet. The combination of the worn diamond plate surface coupled with the accumulation of rainwater was the direct cause of Jack's slip and fall and not his inability to see the water on the floor. The contributing factors to Jack's slip and fall were (a) his preoccupation in performing his job duties, (b) improper lighting, and (c) failure to warn of a wet floor.

Given the fact that Greg had incorrectly identified basic causation information on Jack's incident report may be a reason for why corrective action to improve the safety of the railway's train cars had not progressed. A key element of an incident report includes the collection of good data. Incident reports not only serve as a means of collecting timely information relevant to a guest or employee incident, but also serve as a tool in preventing future incidents. If, however, the direct cause of a particular event is incorrectly coded, as in Jack's case, the incident report will not serve as a warning for future events. It is my view that Greg's error may have been made on previous slip-and-fall incident reports, and appropriate remediation efforts were delayed or not made at all. Such failure to take corrective action (i.e., application of nonskid safety strips) presented an elevated risk of injury to other employees of the railroad, including Jack.

Greg's testimony stated that he believed Jack's slip-and-fall incident was not unique and therefore did not require an announcement in the company's system bulletin, which in turn would have served to alert employees to the potentially dangerous walkway conditions to which action could be taken to alleviate it. This failure on Greg's part to properly identify what was an inherent safety problem exposed all the railroad company's employees to the risk of a slip and fall.

It is clear that a number of preventative measures were not in place that could have prevented Jack's slip and fall. Such measures include proper walkway and stair inspections, the use of slip-resistant flooring and stair treads, appropriate footwear policies, proper hazard identification, proper lighting, appropriate maintenance procedures, and proper employee safety training.

Had the defendant complied with the written maintenance and inspection procedures as described in 49 CFR Chapter II (10-1-99 Edition), Section 238.305 (C)(2), which call for proper and timely inspections by a qualified person, it is likely that the worn and subsequently slippery platform and adjacent steps would have been identified and repaired prior to Jack's slip and fall.

The defendant's failure to maintain, inspect, warn, and remedy the dangerous condition of the platform and stairway in question was, in my opinion, the proximate cause of Jack's slip and fall. The defendant knew or should have known of the industry standards they were obligated to follow to protect the safety of their employees and invited guests. It is my conclusion that the defendant failed to exercise reasonable care to reduce or eliminate the dangerous condition of the platform floor and stairway, and had they done so, Jack's fall would not have occurred.

HOW DID IT END?

The case went to trial in U.S. District Court in Albuquerque, New Mexico. The judge in the case was, like many federal judges, an elderly gentleman in his eighties. My testimony focused on the information I had written in my report, which I must say was a bit lengthy and boring, so much so that midway through my testimony the judge fell asleep. I don't know how long he was sleeping, but it only came to my attention when one of the attorneys objected to a question and the judge did not respond. After what felt like an hour but was only a minute or so, the bailiff walked from the rear of the courtroom, behind the bench, and nudged the judge, who awoke from his slumber. The objection was again made, and the judge replied in a stern voice, "Sustained!" I thought to myself, so this is how the justice system works? Thinking "So what can go wrong?" more did. One of the jurors failed to return after a break, which triggered the bailiff to go and find her. Apparently she thought the trial was over for the day and was headed to her car. What a mess!

The jury ruled in favor of the railroad and based their opinion on two facts. First, that the plaintiff had a preexisting back injury that had nothing to do with his fall, and therefore the railroad company was not responsible, and second, that the railroad company was a big part of the community and was well liked.

43

THE RISING PRICE OF BANANAS!

Sam was just a regular type of middle-aged guy who, one Saturday morning in May of 2012, was summoned by his wife to go food shopping. Sam was the designated cart pusher. As he trailed behind his wife, who was now some fifteen to twenty feet ahead of him in the produce area of the Mega Mart food store in Plano, Texas, he soon came upon a freestanding display of mangoes. The mango display had been built on top of an open-ended wooden pallet. As Sam slowly walked around the corner of the display, he caught the toe of his shoe on the open, uncovered corner of the pallet, which caused him to lose his balance and fall forward, whereby he broke his arm.

It is common for people to trip over retail displays; in this case it was clear that the defendant constructed their mango display as to be in the direct path of pedestrian travel, and in doing so, failed to provide adequate protection to prevent shoppers from catching their foot in the exposed corner(s) of the wooden pallet. Although it is customary for retailers to display palletized products, the primary function of a pallet is to transport material (e.g., produce) from one location to another. Pallets are not retail displays and should not be used as such unless they are properly covered to prevent a pedestrian from engaging their foot between the upper and lower slats of the wooden pallet.

It is customary in the retail industry to cover the open sides of a palletized display to create a toe-kick, which would prevent accidental foot entry. The industry standard for pallet displays calls for the use of a pallet cover or pallet skirt, as depicted in the following figures.

Grocery store retailers like the Mega Mart are aware that if their merchandise displays are improperly maintained or constructed, they can lead to a guest injury.

The large insurers of grocery stores are very much aware of this problem; one of the largest insurers, Zega Insurance Company, publishes a range of information, including a newsletter entitled "Retail Safety Solutions," which specifically addresses this problem and the importance of protecting guests from the dangers of unsafe retail displays. The Zega Insurance report states that "the instability or poor balance of the fixture after assembly, poor weight distribution and/or overloaded weight capacity of the fixture after assembly, improper assembly of the fixture based on manufacturer instructions, empty end caps, poor or no visibility of the base of the fixture, sharp ended display hooks, improper or overloading of merchandise onto the fixture, and damaged components of a fixture."

Furthermore, there are a series of display safety controls that directly affect how retailers design and construct retail displays. Such controls "should include associate procedures and training on the proper assembly and loading of displays and fixtures, safety inspections to include documented repairs, and proper signing of displays such as 'Please Ask for Assistance!'"

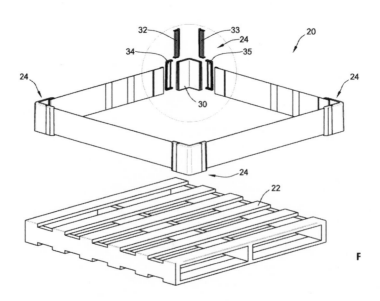

The Zega report further states that "Empty end caps are a common cause of serious trip and fall incidents in the retail industry. Many times, the white or cream-colored end cap shelf blends in with the flooring. As a result, the customer may not see the exposure when turning the corner. In order to prevent these serious injuries, end cap shelves should be maintained full or the empty shelf removed. Also, bulk stack displays should be maintained at three feet in height to prevent customer injuries. Stacks that can no longer be maintained at the appropriate height should be removed from the sales floor."

Secondly, given the fact that the store's management chose to locate the palletized display directly in the walkway where pedestrians are to walk around its corner(s) further enhanced the risk that a pedestrian could catch their foot in the open corner of the pallet. The risk of an accidental foot contact with the display was further enhanced by the fact that the shopper's attention was directed upward toward the merchandise and not down toward the floor.

Lastly, it is unclear if the Mega Mart provided any safety training to their employees. Based on the deposition testimony of the Mega Mart store manager, Fred, he could not recall any policy nor did he recall receiving any training related to the safe maintenance and construction of merchandise displays. Although the Mega Mart had an associate handbook, the book did not contain any information related to mer-

chandising of display safety nor any responsibility on the part of their employees to safeguard the public from an unsafe product display. In short, although the workers at the Mega Mart were instructed to construct floor displays, they were never trained as to how and where to make such displays.

It was my conclusion that Sam's trip-and-fall event could have been prevented had the defendant properly covered the display pallet with a pallet skirt or cover as seen in previous figures in this chapter. The use of pallet skirts and covers is the standard practice in the retail industry and when used, will protect pedestrians from the risk of an accidental trip-and-fall event.

The defendant knew or should have known how to safely construct their merchandise displays and in turn failed to exercise reasonable care as it relates to the specific display in question; therefore, they were directly responsible for Sam's trip-and-fall event and subsequent injuries.

Again, the hazard was created by the Mega Mart. Had they simply trained their workers as to the proper use of pallets as a display element, Sam would not have tripped over the pallet. Furthermore, like many cases I have been retained on as an expert, the retail store's manager didn't check to see if his workers built the display properly.

HOW DID IT END?

The insurance company for the Mega Mart settled with Sam for $180,000. The mentality of many retailers is "Of course we have slips, trips, and falls . . . that's why we have insurance."

PERSONAL NOTE

I once was retained as defense expert for a large grocery store chain in Florida, where after winning a defense verdict, the company executive approached me and thanked me for "winning their case." I replied, "If you called spending nearly $100,000 in legal expenses (not including my fee) winning, what do you consider losing?" He educated me on the economic theory his company had when it came to paying for what he called "slip-down cases."

"You see, Russ," the executive told me, "we look at these slip-down cases as a cost of doing business, kinda like paying the electric bill. No, we don't like it when the electric bill goes up, but what can we do but pay it? Do you know what the highest-selling product (by volume) we sell is?" I answered no. "Well, it's bananas! We sell more bananas than any other single item, and the cost of bananas is flexible. When people come to our store to buy bananas, it really doesn't matter if they are a few cents more or less than our competitors'. So when we have to pay the cost of a slip-down, we simply charge a few cents more per pound for bananas until the claim cost is covered."

Ever since learning this business model, I have found myself monitoring the cost of bananas at my local grocery store. The following photos were taken at my local grocery store a few weeks apart. I guess they had a "slip-down" claim in between the times when I took my photos!

44

FALLS AREN'T FUNNY!

In the fall of 2007 I was retained by an attorney representing a middle-aged woman named Laura, who at approximately 5:30 p.m. on the evening of April 5, 2005, slipped and fell as she exited a Dallas, Texas, grocery store. Laura had gone to the Backyard Grocery Store's pharmacy to pick up a prescription for her daughter. As she was exiting the store, she slipped and fell on a floor that was wet from migrating rainwater. When she fell, Laura struck the back of her head on the polished vinyl composition tiled (VCT) floor, causing her to lose consciousness. Laura, now a bit dazed, was able to drive herself home but later that night began to feel nauseous. Her dizziness had evolved into a throbbing headache, which prompted her daughter to take her to a local hospital emergency room. Once at the hospital, Laura's condition worsened, and she was diagnosed as having broken a series of blood vessels in her brain. The result was permanent brain damage.

The facts of the case were:

1. The exit way floor was wet due to rainwater being tracked in and/or blowing into the doorway.
2. Laura had slipped and fallen on the wet tiled floor located approximately one to two feet from the doorway.
3. The section of floor that Laura slipped and fell on was not carpeted.
4. Store management and employees were aware of the hazardous condition of the floor prior to, and on the day of, Laura's event.

5. Wet floor signs were not posted at the location of the hazardous condition.

Shortly after Laura regained consciousness, she was approached by a store employee by the name of Eric, who completed an incident report, a fact the first attorney never knew. According to the incident report, there were two individuals who actually witnessed Laura's fall, including a gentleman by the name of Brent and a second by the name of Howard. There were several discrepancies in Brent's and Howard's statements. Although everyone agreed that Laura did in fact slip and fall on a wet floor, they did not agree as to a few simple facts. One fact that stood out as being in conflict was that of the presence of floor mats. Eric stated that there were two floor mats in the exit vestibule at the time of Laura's fall; however, both Brent and Howard testified that there were no mats in the area.

It is my opinion that multiple failures had occurred that could have prevented Laura's slip-and-fall event and subsequent injuries. First, it is my view that the management of the grocery store in question exercised poor judgment in their decision to not provide adequate floor matting throughout the vestibule area. Eric stated in his deposition that the mats were not pulled all the way to the door because "if we do that, it will stop the door from closing." Such a statement confirms that Eric was aware of a known hazard prior to Laura's fall and that such a hazard could have been corrected by either (a) removing the tiled floor and installing a recessed carpet mat or carpet tile, or (b) modifying the door to provide ample clearance for the existing floor mat. The failure to address this known safety hazard jeopardized the safety of those pedestrians who enter and exit the store during inclement weather.

Slip-and-fall injuries are a leading cause of accidental injuries and fatalities in America. According to the National Safety Council, falls are the leading cause of emergency room visits and leading cause of accidental fatalities to individuals over the age of eighty-five. According to the National Floor Safety Institute (NFSI), most slips and falls are the result of an unsafe floor, which includes floors that are wet.

Slip-and-fall accidents are the leading cause of guest injuries for the grocery store industry as well, of which grocers are well aware. The defendant's incident records describe numerous previous slip-and-fall incidents in the very same vestibule where Laura's event occurred, but they chose not to address the entry and exit way walkway hazards.

It was my opinion that the defendant failed in their duty to exercise reasonable care as it relates to proper safety precautions in the vestibule and in failing to do such, exposed Laura to a preventable injury. This opinion is supported by several industry standards, including the American Society of Testing and Materials (ASTM) F-1637-02 standard entitled "Standard Practice for Safe Walking Surfaces," which was in effect on the date of Laura's slip and fall and outlines the industry standard of care as it relates to the use of floor mats.

Section 5.1.4 states that "Interior walkways that are not slip resistant when wet shall be maintained dry during periods of pedestrian use." Section 5.4.1 states that "Mats, runners, or other means of ensuring that building entrances and interior walkways are kept dry shall be provided, as needed, during inclement weather. Replacement of mats or runners may be necessary when they become saturated." Section 5.4.4 states that "Mats shall be of sufficient design, area, and placement to control tracking of contaminants into buildings. Safe practice requires that mats be installed and maintained to avoid tracking water off the last mat onto floor surfaces."

Both Brent and Howard reported seeing "dirty water" being tracked onto the tile, thus indicating that there was either insufficient floor matting (i.e., saturated mats) or no mats at all in those areas.

Section 5.4.5 states that "Mats, runners, and area rugs shall be provided with safe transition from adjacent surfaces and shall be fixed in place or provided with slip resistant backing." If in fact there was a floor mat at the exit vestibule, it did not provide for a safe transition to the tiled floor.

It is my opinion that the lack of the floor mats, or the position and/or saturation of the floor mats, located at the defendant's grocery store were in violation of the above-named sections of the ASTM F-1637-02 standard.

The American National Standards Institute (ANSI) has a published standard A1264.2-2001 entitled "American National Standard for the Provision of Slip Resistance on Walking/Working Surfaces," which was in effect at the time of Laura's slip and fall. The standard points out the importance of protecting pedestrians from the hazards presented by floor mats as well as the importance of proper housekeeping, supervision, and warnings. Listed below are the standards as defined by ANSI that the grocery store industry is to follow as related to proper placement and maintenance of floor mats, housekeeping, supervision, and warnings.

Section 6 "Mats and Runners" requires that "Mats and runners shall be adequately secured against movement."

Section E6.2: "Unsecured mats and runners, or mats and runners without slip-resistant backing can be hazardous to users. Taping of mats represents an inexpensive control, which will not only prevent the edges from curling but will also prevent the mat from sliding."

Section 7 "Housekeeping," Subsection 7.1 "General" states that "A housekeeping program shall be implemented to maintain safe walking-working surfaces." Section E7.1: "A written housekeeping program is recommended to ensure consistency and quality. The program should describe materials, equipment, scheduling, methods, and training of those conducting housekeeping." Section 7.4 "Supervision" requires that "The housekeeping conditions shall be monitored and a person(s) shall be authorized to promptly initiate corrective action(s). Section 7.4.1 Monitoring of areas shall include:

a. Inspecting all walking surfaces;
b. Promptly notifying persons responsible for clean-up of affected conditions;
c. Placing signage, barriers or personnel until clean-up is complete."

Section 8 "Warnings," Subsection 8.1 "General" states that "a warning shall be provided whenever a slip/fall hazard has been identified until appropriate corrections can be effected." Section E7.4 states that "Good housekeeping should require all employees to identify and report potential hazards to appropriate supervision." Section 8.1.1 states that "When a slip/fall hazard which covers an entire walkway exists, thus making it difficult to safely route personnel around the hazard, barricades should be used to prevent access (see Section 9.1). If appropriate, an employee should be assigned to detour personnel, in conjunction with the appropriate use of warning signs until the barricade can be erected or the hazard removed."

It is my opinion that the defendant failed to meet the industry standard listed above and in doing so, allowed a known wet floor safety hazard to exist, thus exposing their customers and employees to a potential slip-and-fall event.

It is also the industry standard to immediately correct any known hazard once it has been identified, which is echoed in the defendant's employee training and safety policies. The section on prevention of slip/fall accidents as stated on page 5 of the defendant "Play It Safe" safety guidelines requires that "All employees have the personal responsibility to prevent slip/fall accidents" and that when confronted with a floor safety hazard, employees are required to "warn fellow employees and customers of the floor hazard." Although there were numerous store employees in the area at the time of Laura's fall, including the store manager, none of them verbally warned Laura as to the impending slip hazard that was known to be present. Such failure to warn was in direct violation of the defendant's safety policies.

Based on this information, it is clear that the management and staff of the defendant's store had in fact allowed a known slip-and-fall hazard to exist, which led to Laura's slip-and-fall event and related injuries. Had the defendant's management addressed the above-mentioned walkway hazards prior to Laura's visit, it is unlikely that she would have slipped and fallen.

It is therefore my conclusion that the defendant store in question was grossly negligent, reckless, and negligent in their duty to protect their invited guests from unnecessary harm and was directly responsible for Laura's injuries.

During the discovery stage dozens of depositions were taken, including numerous store employees'. One document that the defendant would not produce was a list of previous slip-and-fall events. It finally took a threat of contempt by the judge to compel the defendant to produce their previous incident reports. Twelve incident reports were produced, all of which were women who had slipped and fallen at the same store within the past five years. Each was signed by the customer and contained language stating that their slip and fall was their own fault and that they were simply "not watching where they were walking."

THE TRIAL

After nearly five years of waiting Laura had her day in court. Prior to trial I was asked to give a deposition. The defendant's attorney was a woman by the name of Patricia, who to this day holds the record as the worst defense attorney I have ever met. She was rude and obnoxious throughout my six-hour-long deposition and simply set out to draw me into a string of arguments she had set out as a trap. The more she tried to argue with me, the more I steered her back to the factual basis of my opinions. Needless to say, this was a deposition from hell. I later learned that Patricia had done the same to Laura, who at the time of her deposition was extremely weak and fragile. The mistake Patricia made was to videotape Laura's deposition, which was played to the members of the jury at trial. Patricia's mean-spirited attack on Laura did not play well with the jury, who were sympathetic to Laura.

At trial a number of the defendant's employees were called to testify, including their corporate risk manager, Peter. Peter was the consummate corporate kiss-ass, whose arrogant and condescending personality further hurt his employer's case. When asked, "How many slip-and-fall incidents were you aware of in your twenty-plus-year career as the risk manager for the Backyard Grocery Store chain?" Peter answered, "Oh, at least one thousand," which drew a follow-up question of, "Of those thousand previous slip-and-fall incidents, how many were the fault of the grocery store?" Arrogant Peter answered, "None!" What a ridiculous answer. It was clear that Peter and Eric thought that they would outsmart the jury and worked out their testimony in advance. A bad idea indeed!

At the time of trial six of the twelve women who had completed an incident report for a slip and fall at the defendant's store were called to testify. The remaining six women could not be located. When asked if they recalled their slip-and-fall event, all of them said yes, but that they did not recall completing an incident report. When asked by John whether or not the signature at the bottom of the report was theirs, they each answered, "No." What was even more interesting was that all of their signatures looked eerily similar, and they may have been authored by a single person.

At trial the defendant called three Hispanic employees and claimed that they were all witnesses to support the fact as documented by Eric that floor mats were in fact placed at the door where Laura slipped and fell. However, when each of the gentlemen was asked as to the color of the mat, they all gave different colors. After a bit of pressure, one of the employees paused, looked down at his lap, and said, "There were no mats at the door. I was told to say that there were and if I didn't I would be fired." When John asked the worker who threatened him, he said, "That man over there," and pointed to Peter. The attorney for the Backyard Grocery Store quickly called for a sidebar, whereby the jury was asked to leave the courtroom. Strike one against the defendant.

Once the jury had reconvened, Peter was asked to take the stand a second time and explain himself. Surprisingly, he began to weep, telling the jury that he felt it was his responsibility to protect his employer and the employees of the grocery store and that if Laura was to prevail in her case, the cost would bankrupt the company. He claimed that "the jobs lost by this lawsuit would hurt the community which the Backyard Grocery Store chain served." When asked if he had ever completed or modified an incident report, he answered, "Yes," and later acknowledged that he often authored slip-and-fall incident reports and signed them in lieu of the guest. Strike two!

My testimony was clear and to the point. The defendant failed to provide a safe exit for their invited guests. If they had they done so, Laura would not have been injured. My testimony concluded with a question regarding the protocol of a risk manager to complete and sign incident reports on behalf of a customer. I answered that such procedure was improper and possibly criminal. When asked what I thought of Peter's admission of falsely completing such reports, I said in a clear voice, "He belongs in jail." By this time it was late in the day, and the jury was dismissed for the day. Before I left the building I made a stop in the men's restroom, and none other than Peter followed me in. To my surprise Peter uttered, "It's easy for you to criticize me; you never had the responsibility of running a grocery store." I replied, "You're right. But I sure as hell would do a better job than you!"

HOW DID IT END?

After an exhausting two-week trial the jury found in favor of Laura, awarding her $4.5 million, the largest verdict in the State of Texas's history for a slip and fall. Most of the award was for punitive damages. Punitive damages are very rare, especially in a slip-and-fall case, and reflect anger on the part of the jury toward the defendant. Punitive damages are the jury's way to send a message to a defendant that they hope will cause them to change their behavior. The jury was pissed and so was I. So much so that it inspired me to write my second book, entitled *Falls Aren't Funny*. A year or so later, a made-for-TV episode by the same name was produced.

Sadly, because of the threat of an appeal, Laura received a much lesser amount. For those who will naturally think that $4.5 million is a lot of money, I would ask you to consider the following. Laura is permanently disabled and cannot care for herself. The average cost for Laura's nursing care is $5,000 per month or $60,000 annually. Given Laura's age and expected longevity, the cost to simply care for her will top $3 million. Would you trade places with Laura?

On a personal note, I thank God Laura found John, who was her unsung hero. John and his brother Mike, who tried the case, were true professionals: quiet and soft-spoken gentlemen who commanded the jurors' respect. John and Mike invested years of their time and hundreds of thousands of their money to represent Laura, all of which would have been a personal loss to them. John is no longer in Dallas but now practices law with Mike and his brother Bill in the Midwest. The Backyard Grocery Store chain went out of business, selling their stores to independent owners.

45

ALASKAN ICE!

On the morning of April 10, 2012, a young mother by the name of Crystal had just dropped her children off at their Nome, Alaska, elementary school and went on to work. About an hour or so later the school called to tell her that one of her children was ill and needed to go home. Crystal promptly returned to the school. It was approximately 10:00 a.m. when she arrived at the school. Crystal parked her SUV in the school's parking lot and stepped out of the car. It only took a few steps until she slipped and fell on a layer of densely packed snow and ice, striking the back of her head.

The accumulated snow/ice was not properly removed or treated with an appropriate deicing compound as to create a uniformly safe walking surface. Warning signs were not posted alerting the public to the impending slip hazard. At the time of her fall Crystal was wearing a pair of blue-colored Dansko shoes and was walking at a normal pace.

It is my further understanding that on the morning of April 10, 2012, the defendant was aware of the icy conditions that existed on their parking lot and therefore was on notice of the dangerous condition on the premises. After all, this was a typical winter day in Alaska! Snow and ice are common. For the defendant to not have a preplanned snow removal process in place is in and of itself a violation of the standard of care.

Subsection 5.1.3 of the American Society of Testing and Materials (ASTM) F-1637-09 *Standard Practice for Safe Walking Surfaces* requires that "Walkway surfaces shall be slip resistant under expected environmental conditions and use. Painted walkways shall contain an abrasive additive, cross cut grooving, texturing or other appropriate means to render the surface slip resistant where wet conditions may be reasonably foreseeable."

Section 5.7 "Exterior Walkways," Subsection 5.7.1 requires that "Exterior walkways shall be maintained so as to provide safe walking conditions." Subsection 5.7.1.1 requires that "Exterior walkways shall be slip resistant."

Subsection 5.7.1.2 requires that "Exterior walkway conditions that may be considered Sub-standard and in need of repair include conditions in which the pavement is broken, depressed, raised, undermined, slippery, uneven, or cracked to the extent that pieces may be readily removed."

Subsection 7.1 "General" of the American National Standards Institute (ANSI) A-1264.2-2006 Standard for the Provision of Slip Resistance on Walking/Working Surfaces requires that "A housekeeping program shall be implemented to maintain safe walking-working surfaces. (E7.1 A written housekeeping program is recommended to ensure consistency and quality. The program should describe materials, equipment, scheduling, methods, and training of those conducting housekeeping.)"

Section 7.2 "Maintenance Procedures" states that "If there are written procedures, they shall specify cleaning and maintenance procedures including immediate response, routine operations, and remedial measures and reporting requirements. (E7.2 Procedures should be reviewed regularly and updated as needed so that an effective program is maintained. Drains should be kept clear and free flowing. Certain spills involving hazardous chemicals may be subject to regulatory reporting.)"

Section 8 "Warnings/Barricades," Subsection 8.1 "General" requires that "A warning shall be provided whenever a slip or trip hazard has been identified until appropriate corrections can be made, or the area barricaded. (E8.1 Slip/trip hazards should be reduced by design, maintenance and/or layout of the work area.)"

Subsection 8.1.1 "Alternate Route" states that "When there is a slip/fall hazard, which covers an entire walkway, thus making it difficult to safely route personnel around the hazard, barricades shall be used to limit access (see Section 9.1). If appropriate, assign an employee to detour personnel, in conjunction with the appropriate use of warning signs until the barricade can be erected or the hazard removed."

Section 10.3 requires that "Snow removal efforts shall be performed as expeditiously as conditions and resources allow." Subsection 10.3.1 further requires that "Where snow and ice exists in pedestrian walkways safe maintenance techniques shall include plowing, shoveling, deicing, salting, and sanding, as needed. (E10.3.1 Chemicals can be used to assist melting of snow and ice and should be applied to pavement and sidewalks, as needed.)"

Subsection 10.3.2 "Responsibilities" requires that property owners "establish the responsibilities for facility managers, custodians, grounds maintenance staff, and contracted snow removal personnel. (E10.3.2 Effective snow management is anticipatory. Weather forecasts should be monitored. It is advisable to establish a threshold over which removal operations should commence [e.g., more than one inch of snow accumulation, and/or sleet and iced-over conditions].)"

Subsection 10.3.3 "Resources" requires that "Appropriate training shall be provided for maintaining safe walkway surfaces. The location of equipment and supplies shall be known to those who will use them. (E10.3.3 Equipment and supplies should be available and ready for use prior to the start of the cold weather season.)"

Subsection 10.3.4 "Priorities" requires that "Priorities for removal shall be established. (E10.3.4 For example, a plan to prioritize snow removal might look like this:

1. Fire lanes must be open for emergency equipment. Fire hydrants must be free of snow and accessible at all times.
2. Main entrances, ADA ramps and curb cuts, weather exposed stairs, and primary sidewalks and parking lots should be cleaned before the building opens.
3. Parking lots, secondary entrances, and other low-usage areas should be cleaned by noon.)"

It was my opinion that the defendant had failed to meet and was not in compliance with the above-named nationally recognized industry standards, which serve as the basis of establishing reasonable care of parking lots specifically as it relates to the prevention of ice- and snow-related slips and falls. Furthermore, multiple failures on the part of the defendant had occurred that could have prevented Crystal's slip and fall and subsequent injuries. First, the defendant's employees were aware that a layer of snow/ice was present from a previous storm and that the presence of ice on their parking lot pavement created a serious slip hazard. Given such, it is my opinion that the defendant failed to provide reasonable care by not properly removing the snow/ice hazard that in my opinion was the proximate cause of Crystal's fall and subsequent injuries.

It is my understanding that the defendant did not have an ice/snow removal plan or policy but relied solely upon the courtesy of the DOT, whom he was aware did not always provide timely snow/ice removal. The school district's head of maintenance was a gentleman by the name of Phil who stated in his deposition that the "DOT was the only one in town that could spread sand and actually had sand to spread" and that "normally we don't get a visit from them until maybe 10:00 and they've got a lot of other areas to cover too. The standard would be 10:00, 11:00 in the morning before they get to us. And from what I understand they got there shortly after the accident." Phil had limited training in slip-and-fall prevention and could only recall watching a thirty-five-minute online course on proper snow and ice removal.

Due to a previous snowstorm, the defendant knew or should have known that the parking lot of the Nome Elementary School would have been hazardous and therefore dangerous to anyone who might park there and seek to access the building's entrance. Therefore the defendant owed a duty to their invited guests, specifically Crystal, to maintain the premises in a manner that made it safe for invited guests to enter and exit the building. Furthermore, if a walkway hazard is recognized, the defendant has the duty to warn their invited guests and invitees as to any unsafe conditions on the premises. It is my understanding that the defendant did not post any type of warning in the parking lot prior to or at the time of Crystal's slip-and-fall event. It is my opinion that the defendant breached these duties and that such failure was the proximate cause of Crystal's fall and subsequent injuries.

Had the defendants (1) properly applied an appropriate deicing compound equally along the surface of the parking lot walkway in question, (2) removed the accumulated buildup of snow/ice by way of shoveling, and (3) properly posted warning signs alerting their invited guests as to the impending slip hazard(s) that existed on the premises, it is unlikely that Crystal would have slipped and fallen. It is my concluding opinion that the defendants were negligent in failing to provide reasonable care as it relates to providing a safe entrance to their invited guests, specifically Crystal.

HOW DID IT END?

The school district settled with Crystal for $90,000.

46

FELL AT THE WELL!

Gabriel was a thirty-year-old driver for a North Texas oil and gas company who on the morning of March 24, 2012, fell down an industrial staircase at a natural gas tank battery located in a small Texas town. If you are not familiar with gas and oil wells, they are generally in remote locations. So remote, most people have never even seen one, even though there are thousands of such facilities across the State of Texas.

Gabriel's job was to drive his water truck out to the natural gas facility to keep the water tanks full. After pulling his truck up to a ten-thousand-gallon water tank, Gabriel got out of his truck and connected his truck's hose to the tank to begin the filling process. He then walked over an elevated steel platform that provided access to the back side of the tank, where the water valve is located, to open it for filling. As he walked back over the steel platform, he slipped off one of the metal stair treads, striking his head on the steel stair below and rendering him unconscious. Gabriel was unable to move his lower body and remained on the ground the entire day. Realizing that something was wrong, Gabriel's employer began to backtrack Gabriel's route, eventually finding him some twelve hours later. Gabriel was transported to the hospital, where he was treated for a neck fracture.

Section 5 "Walkway Surfaces," Subsection 5.1.3 of the American Society of Testing and Materials (ASTM) F-1637-09 entitled "Standard Practice for Safe Walking Surfaces" states that "Walkway surfaces shall be slip resistant under expected environmental conditions and use. Painted walkways shall contain an abrasive additive, cross cut grooving, texturing or other appropriate means to render the surface slip resistant where wet conditions may be reasonably foreseeable."

Furthermore, insurance companies including Liberty Mutual publish information related to painted walking surfaces. Liberty Mutual Insurance recommends a range of aggregate material sizes to be used for various painted walkway applications. An exterior fixed industrial stair that is exposed to a wide range of contaminants would require an aggregate size of two to ten.

Section 5.7 "Exterior Walkways," Subsection 5.7.1 requires that "Exterior walkways shall be maintained so as to provide safe walking conditions." Subsection 5.7.1.1 states that "Exterior walkways shall be slip resistant." Subsection 5.7.1.2 states that "Exterior walkway conditions that may be considered substandard and in need of repair include conditions in which the pavement is broken, depressed, raised, undermined, slippery, uneven, or cracked to the extent that pieces may be readily removed."

Section 7.1.2 requires that "Step nosings shall be readily discernible, slip resistant, and adequately demarcated. Random, pictorial, floral, or geometric designs are examples of design elements that can camouflage a step nosing."

This requirement is echoed in the Occupational Health and Safety Administration (OSHA) CFR 1910.24, Section (f), which requires that "All treads be reasonably slip resistant and the nosings shall be of non-slip finish." The nosings on the stairway in question were not in compliance with this standard.

Section 11 "Warnings," Subsection 11.1 requires that "The use of visual cues such as warnings, accent lighting, handrails, contrast painting, and other cues to improve the safety of walkway transitions are recognized as effective controls in some applications. However, such cues or warnings do not necessarily negate the need for safe design and construction."

Subsection 11.2 states that "When relying on applications of color as a warning, provide colors and patterns that provide conspicuous markings for the conditions being delineated, their surroundings, and the environment in which they will be viewed by users. Bright yellow is a commonly used color for alerting users of the presence of certain walkway conditions. When properly applied, and maintained, other colors can also provide effective warnings."

It is the generally accepted safety practice to post caution signs adjacent to areas where slippery walkways and/or stairs may present a slip hazard. It is my understanding that caution signs were not posted at or near the staircase, and therefore Gabriel was not properly warned as to the impending slip-and-fall hazard created by the wet/slippery steps. The need to properly post warning signs is required under the U.S. Code of Federal Regulations CFR 29 Section 1910.145.

Section 7 "Housekeeping," Subsection 7.1 "General" of the ANSI A1264.2-2006 standard "Provision of Slip Resistance on Walking/Working Surfaces" requires that "A housekeeping program shall be implemented to maintain safe walking-working surfaces."

According to E7.1, "A written housekeeping program is recommended to ensure consistency and quality. The program should describe materials, equipment, scheduling, methods, and training of those conducting housekeeping."

In Subsection 7.4 "Supervision," "The housekeeping conditions shall be monitored and a person(s) shall be authorized to promptly initiate corrective action(s)." E7.4: "An effective program requires all employees to identify and report potential hazards to appropriate supervision. Documentation of monitoring can assist in identifying hazards (e.g., areas where repeated spills occur), which will permit better planning and anticipation of such events."

Section 7.4.1 Monitoring says that "Monitoring of areas shall include:

a. Inspecting all walking surfaces.
b. Arranging for prompt notification of persons responsible for clean up of hazardous conditions."

Section 8 "Warnings," Subsection 8.1 "General" states that "A warning shall be provided whenever a slip/fall hazard has been identified until appropriate corrections can be effected."

Section 10.1 "Safely Maintained" requires that "Walking surfaces for use in accordance with 2.3 shall be safely maintained." Section 10.2 "Practical Considerations" states that "Where it is not practical to replace flooring, etching, scoring, grooving, brushing, appliqués, coatings and other such techniques shall be used to provide acceptable slip resistance under foreseeable conditions." Section E10.2: "Surfacing applications and/or treatments are available that can impart increased slip resistance to problem surfaces. Some flooring surfaces can have their surface traction enhanced by etching. Certain paint on or trowel on applications can enhance slip resistance. It is important to select one that will adhere tenaciously to the substrate. Cleanability and durability should be considered. Patch testing of prospective materials in a problem environment is recommended before proceeding with general application. Carpeting is also an option worthy of consideration for control of slips."

OCCUPATIONAL SAFETY AND HEALTH ADMINISTRATION (OSHA)

Code of Federal Regulations Subpart D. Section 1910.24(f) "Fixed Industrial Stairs" requires: "Stair treads: All treads shall be reasonably slip-resistant and the nosings shall be of nonslip finish. Welded bar grating treads without nosings are acceptable providing the leading edge can be readily identified by personnel descending the stairway and provided the tread is serrated or is of definite nonslip design. Rise height and tread width shall be uniform throughout any flight of stairs including any foundation structure used as one or more treads of the stairs."

OSHA further requires that employers like the defendants complete a Job Safety Analysis (JSA), which requires that workplaces be inspected for potential hazards. According to OSHA: "To complete a JSA effectively, you must identify the hazards or potential hazards associated with each step. Every possible source of energy must be identified. It is very important to look at the entire environment to determine every conceivable hazard that might exist. Hazards contribute to accidents and injuries." It is unknown whether or not the defendants completed an OSHA-required JSA.

It was my opinion that the defendant was not in compliance with the industry safety standards that comprehensively establish the standard of care for industrial work areas.

Slips and falls resulting from wet floor hazards are a leading cause of employee injury for the oil and gas industry. For this reason, most oil and gas companies have adopted a rigorous walkway inspection and maintenance program as well as a comprehensive employee-training program. It is unclear as to the defendant's policies regarding walkway safety, walkway inspections, as well as employee safety training guidelines; however, based on the defendant's employee's deposition testimony, it is clear that such policies and procedures were either nonexistent or inadequate.

The oil and gas company knew that steps at their tank batteries would become slippery from rain, mud, or other contaminants and therefore had both an opportunity and obligation to properly maintain the safety of their industrial stairways under such conditions.

The company had painted the steps but failed to use an appropriate granular additive as to render them slip resistant. By improperly painting the steps the company reduced the built-in slip resistance, which was created by the sharp, edged perforated circular pattern, punched into each stair by the metal fabricating manufacturer. By failing to use the proper type of paint, which was absent an antislip additive, the company created a condition of reasonable harm to anyone using the stairs, specifically Gabriel.

Based on the facts as presented to me at that time, it was my opinion that Gabriel's slip-and-fall event was predictable and preventable, and had the defendant (1) painted the steel steps as to use an appropriate aggregate material, (2) warned of a potential slip hazard via the use of a posted warning sign(s), or (3) used a more appropriate stair tread material, one that would provide superior slip resistance when wet, Gabriel would not have been exposed to a stairway hazard and subsequently slipped and fallen. It is my view that the failure to respond to the hazard is most likely a combination of improper employee safety training, vague maintenance and inspection policies, and the lack of proper maintenance (i.e., appropriate paint and aggregate) on the part of the company's safety and/or maintenance staff.

It is my further opinion that the company had failed in their responsibility to provide a safe walking surface for their workers and contractors, specifically Gabriel, and in doing so, created an unreasonably dangerous condition. Had the company been in compliance with the industry safety and maintenance standards, which include proper hazard removal, the proper posting of hazard warnings, providing slip-resistant stair steps and nosings, and walkway inspection procedures, it is unlikely that Gabriel would have slipped and fallen.

It was my opinion that the defendants did not exercise reasonable care as it relates to the safety of their walkways and stairways and therefore were negligent in their duty to protect their workers, specifically Gabriel, from the risk of injury. Had the company exercised reasonable care as it relates to their walkways, it is unlikely that Gabriel would have slipped and fallen.

HOW DID IT END?

Gabriel settled with the oil and gas company for $650,000.

47

HURT AT THE HOSPITAL!

In 2011 I was retained by an anesthesiologist by the name of Dr. Cabernet, who caught the heel of his shoe on a floor mat placed in front of an elevator, causing him to fall backward and strike his head on a nearby steel-framed sofa. Dr. Cabernet's fall resulted in him severing his spinal cord in his neck, making him a quadriplegic for life. The event was captured on the hospital's surveillance and was a key piece of evidence.

Dr. Cabernet sued the hospital and the hospital's contracted elevator company, who was working on the elevator at the time of his fall, charging that they were negligent in that they created the mat defect that caused his fall and subsequent permanent injury.

After performing a site inspection of the hospital, it was my opinion that the floor mat in question presented an unreasonably dangerous condition that is outlined in the plaintiff's supplemental answers listed below.

HOW DID IT END?

The elevator company settled with Dr. Cabernet, which left the hospital as the last standing defendant. The case was discharged by the Cook County judge and reversed in part by the State of Illinois, First District, Fifth Division court. Listed below is a case overview as published by the State of Illinois Official Reports outlining a case.

Dr. Cabernet took his case all the way to the State of Illinois Supreme Court, which reversed the lower court's decisions. An overview of their findings was published in the *McKenna Law Update & News*.

Dr. Cabernet settled the case for an undisclosed eight-figure sum. In 2014 the local PBS station did a "Life in the Day" program that showed the struggle he and his family were enduring as a result of his fall. The program also revealed the courage, strength, and dignity he and his family had in dealing with their crisis, which they managed gracefully. Dr. Cabernet passed away on Sunday, June 29, 2015. He was seventy-one years old.

48

FIRST-CLASS DELIVERY!

In 2014 I was retained by a woman named Marylyn, who a few days before Christmas tripped over a curled and broken edge on a clear plastic logo-type mat at the post office. Marylyn's fall resulted in serious injury to her wrist, arm, and shoulder.

After reviewing the case file that included examining the mat in question, I prepared a written report. My opinions were based in part on the appropriate industry standards, which included the American Society of Testing and Materials (ASTM) F-1637 standard entitled "Standard Practice for Safe Walking Surfaces," which establishes the standard of care by which property owners shall maintain, inspect, and replace floor mats like that in this case. A second industry standard was also in violation: Section 6, entitled "Mats and Runners," of the American National Standards Institute (ANSI) A1264.2 standard.

Last was a reference to the Americans with Disabilities Act (ADA) 2010. Section 303 of the ADA entitled "Changes in Level," Subsection 303.2 "Vertical" requires that "Changes in level of 1/4 inch (6.4 mm) high maximum shall be permitted to be vertical."

Section 304.2 "Floor or Ground Surfaces" requires that "Floor or ground surfaces of a turning space shall comply with 302. Changes in level are not permitted."

Advisory 304.2 "Floor or Ground Surface Exception" states that "As used in this section, the phrase 'changes in level' refers to surfaces with slopes and to surfaces with abrupt rise exceeding that permitted in Section 303.3. Such changes in level are prohibited in required clear

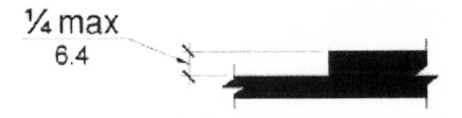

floor and ground spaces, turning spaces, and in similar spaces where people using wheelchairs and other mobility devices must park their mobility aids such as in wheelchair spaces, or maneuver to use elements such as at doors, fixtures, and telephones. The exception permits slopes not steeper than 1:48."

My opinions were also based on the deposition testimony of the post office's supervisor and maintenance man. The post office's customer service supervisor stated in her deposition testimony that "there were no defects or hazardous material" associated with Marylyn's trip-and-fall event and that "the lobby is routinely inspected and that usually a maintenance person goes out every morning between 7:00 and 7:30 to make sure the place is clean." The supervisor stated that she was aware that a customer had witnessed the plaintiff's trip and fall and that she "had a written statement from the witness in her file" and that "she did not remember speaking to any witnesses that led her to believe that the statement by Marylyn was incorrect or wrong." The supervisor completed a Personal Injury Incident report and documented that Marylyn had somehow caused her own injury by some form of "unsafe practices, inattention or distraction." The supervisor further suggested in the post office's incident report that Marylyn's footwear was the cause of her fall. There is absolutely no evidence or eyewitness statements to support either claim.

The supervisor further stated that the post office's "maintenance personnel were responsible for placement of the advertising mat" in question. On the day in question the maintenance person was Ronald. The supervisor further stated that after Marylyn's fall, she had inspected the mat in question and found that "it was straight, dry, nothing wrong with the mat" and that "she did not consider it a safety concern." However, she then made the decision to remove the mat from service.

The supervisor stated that she, along with two other employees, serve on the post office's safety committee, which meets quarterly and addresses such topics as "anything related to employee safety or customer safety." However, she was "unsure" whether or not Marylyn's trip-and-fall incident, which she recorded, was ever discussed by the committee. The supervisor also stated that maintenance personnel's level of safety training is limited to "reading material," and she was unsure if maintenance personnel had any ongoing safety training.

In Ronald's deposition, he stated that he would physically check the mat in question "probably once a month" for the purpose of "looking for wear." Had he noticed any defects or problems with the back of the mat, he would have "pulled the mat and ordered a new one."

Ronald did not recall the mat in question as having any tab(s) and did not recall reading any instructions as to the proper use and maintenance of the clear plastic mat. Ronald recalled being asked by the supervisor as to his opinion of the mat. He inspected it and "flipped it over and looked at all the rubber and the rubber was still soft, not bent." He also "got down on his knees to look at the rubber and check to see if anything like maybe the ends were cracked or something. He claims to have seen nothing."

It is my opinion that the mat in question did not provide for a safe transition as defined by the nationally recognized industry standards and law listed above. It is also my opinion that the defendant's employees whose responsibilities include walkway safety inspections, specifically Ronald, failed to comply with the United States Postal Service (USPS) published floor safety rules, which specifically require that employees "report defective walks, steps, and parking surfaces so that repairs to eliminate tripping hazards can be made promptly" and to "secure carpets, rugs, and mats and arrange them to prevent slipping. Repair or replace those with wrinkles, turned-up edges, or tears." Section K of the USPS's policy requires postal employees to "give customer areas special consideration they deserve. The improper placement of mats or rugs, or lack of them can result in customer injuries and significant liability to the Postal Service."

Both the supervisor and Ronald stated in their depositions that they had inspected the mat in question immediately after Marylyn's trip-and-fall event and did not see any defects, yet the photographs of the mat in question taken after Marylyn's fall clearly show multiple defects, including a cracked edge, a broken and elevated left corner, and delaminated backing material. How both the supervisor and Ronald did not recognize such significant damage is difficult to comprehend. The floor mat in question clearly shows significant damage, and such failure on the part of the defendant to properly identify through even a cursory inspection such damage either suggests that they (a) did not inspect the mat or (b) did not consider the condition of the mat to be damaged.

Based on the photographs of the clear plastic mat, it appears that the cracked left corner is elevated approximately 1/4 inch to 1/2 inch in height and therefore posed a walkway trip hazard as defined by the nationally recognized safety standards and code listed above. Failure to observe such standards suggests that the defendant failed to provide reasonable care as it relates to the safe use, inspection, and care of the floor mat in question.

It is impossible to know exactly how the corner, edge, and backing of the mat were damaged; however, the manufacturer of the mat instructs the user to "never use the product if damaged" and shows placing the mat with the tabbed corner away from the main traffic route, which the defendant failed to do and because of such, may have contributed to damaging the tabbed corner of the mat.

It is my opinion that failure to inspect and remove the mat in question presented an unreasonable risk of harm to pedestrians, specifically Marylyn, and was the proximate cause of Marylyn's trip and fall and related injuries. By not properly inspecting and removing the damaged and hazardous mat from service the defendant failed to provide a safe walking surface to their invited guests.

Furthermore, there is no evidence to suggest that Marylyn had acted in an unreasonable manner, or that she was not exercising reasonable care as she walked across the mat in question toward the defendant's customer counter. Given such, it is my opinion that the floor mat in question was damaged and presented an unreasonable risk of harm to anyone walking across it.

Had the defendant addressed the above-mentioned walkway hazard prior to Marylyn's visit, and had they taken proper care of the property within its custody and control, it is unlikely that she would have tripped and fallen and subsequently injured herself.

HOW DID IT END?

I was called to testify at trial, which was a bench trial (no jury) at the U.S. Federal Courthouse in Detroit, and was asked a series of questions about the floor mat. I opined that not only was its top layer curled, but it was cracked and broken. The mat's slip-resistant backing had separated from the back of the mat, further compromising its ability to remain stable when walked on. In short the mat in question was severely damaged and defective.

Marylyn lost her case. The judge found in favor of the post office. However, the clear advertising-style floor mats are no longer used by the USPS.

49

BAD SIGN!

On December 24, 2003, Robin tripped and fell as she was leaving the Southlake Theater in Georgia. She caught her foot in the open-handle portion of a wet-floor sign that had folded over and was lying flat on the ground. Robin's daughter Dawn was with her as they exited the movie and witnessed her mother's fall. Robin and her daughter both recalled just how crowded the theater was that day, and they were exiting their movie around the same time other movies were letting out. Robin and Dawn found themselves absorbed into a herd of people all moving toward the theater's front exit. The crowd made it difficult to see the wet-floor sign that was lying on the floor, and before she knew it she was lying flat on the ground.

Based on the deposition testimony of Jorge, the theater's supervisor on duty at the time of Robin's fall, he remembered seeing a spill on the carpeted lobby floor before the movie started, soaked it up with towels, and placed the yellow A-frame-style wet-floor sign near the spill. He further stated that they had about a dozen such signs in the theater.

OPINIONS

It was my opinion that the carpeted lobby in question presented a slip-and-fall risk due to the presence of a wet area where the defendant had properly placed the wet-floor sign. The type of wet-floor sign selected by the defendant's management is one that is prone to collapsing upon contact by pedestrian traffic. This tendency to collapse upon contact makes it inappropriate for use in areas such as the exit lobby in a movie theater. Given that this type of floor sign was in fact used at the time of the plaintiff's fall, the defendant and their management had the responsibility to ensure that the sign be used in a manner that is safe for pedestrian traffic. The sign in question had most likely been knocked down by one of the people in the crowd, thus rendering the sign useless as a warning and in turn created a potential walkway hazard. Had the management and/or the employees of the theater in question used an appropriate sign, one with a larger, more stable base, it is unlikely that the plaintiff would have experienced her fall.

It was also my opinion based on the deposition testimony of the defendant's employees that they were improperly trained in hazard identification and floor sign use. Furthermore, based on the deposition testimony of Robin and Dawn, they recalled seeing two theater employees standing near the scene at the time of the fall, and that they were aware that the wet-floor sign had been knocked down. Jorge testified that if the wet-floor sign were knocked down and lying on the floor, it would still serve as an adequate warning to pedestrians. Such a statement was further proof that the defendant had failed to properly train their employee as related to the proper use and placement of wet-floor signs.

Had the defendant utilized a wet-floor sign that was constructed with a more stable and firm base, it is unlikely that the sign would have been knocked down by pedestrian traffic. Unfortunately, the defendant and their management had chosen to use a sign that was prone to falling and therefore increased the likelihood of a pedestrian tripping and falling on the sign.

It is therefore my opinion that the defendant had failed to provide a safe walking surface for their invited guests, and had they simply removed the sign once it had tipped over or used a sign with a more stable base that was less prone to tipping over, it is unlikely that Robin would have tripped on the sign.

HOW DID IT END?

The defendant was granted a summary judgment by the county court primarily based on the argument that the employees did not actually see the sign on the ground and therefore did not have actual notice. Robin appealed her case and lost at the appellate level, forcing her to either abandon her claim or appeal to the Georgia State Supreme Court, which she did. Surprisingly, based in part on my expert opinions, the high court ruled in favor of Robin and challenged the claim that the theater's workers did not see the sign when their testimony stated otherwise. The high court sent the case back down to the county court to be tried. Surprisingly, Robin's attorney chose not to call me to trial, which sadly resulted in Robin losing. The jury felt that had Robin been looking where she was walking, she would have seen the sign on the ground and avoided stepping on it like all the people ahead of her did.

P.S. The movie Robin and Dawn had gone to watch was a remake of Steve Martin's *Cheaper by the Dozen*.

50

OSHA GETS THE BOOT!

In the summer of 2010 I received a call from a defense attorney in Denver, Colorado, who asked if I would mind doing a telephone interview with him and his partner regarding a case they were working on; I agreed. Although it is common for attorneys and experts to discuss a case prior to retention to ensure that there is no conflict and that the expert does in fact have the expertise for the particular case, this was different. For nearly an hour I was asked a wide range of questions regarding my knowledge of the Occupational Health and Safety Administration, otherwise known as OSHA. I said that I was very familiar with OSHA; however, my expertise is limited to OSHA's Section 1910 on walking and working surfaces. They then asked what my relationship was to OSHA, and I responded that I have received training as it relates to OSHA and am the recipient of a ten-hour and a thirty-hour training course card. I also mentioned that in May of 2006 I had attended the induction ceremony of then Assistant Secretary of Labor for OSHA Ed Foulke, whom I had met during my tenure on the Board of Delegates for the National Safety Council, and that in January of 2011 I was invited to provide testimony at the OSHA hearing in Washington, D.C., regarding OSHA's proposed revision to their walking and working surfaces rule.

They then stated that they had learned that I was one of the leading experts on slip-resistant footwear and wanted to know what my opinions are as related to OSHA's requiring that employers provide slip-resistant shoes to their workers. I responded, "There is no such requirement."

Having worked with Walmart Stores USA and Canada in the mid-1990s on developing one of the nation's first retail line of slip-resistant footwear, I was very aware as to the use of slip-resistant footwear in the workplace. Back in the mid-1990s the Traction Plus-TRED SAFE line of slip-resistant footwear was the number-one best-selling line of slip-resistant shoes to the fast food industry, the hospitality industry, and others. Employers then and now had three ways to get their workers to wear slip-resistant shoes: first and most popular, was to make slip-resistant shoes a part of their job requirement whereby the employer would assist the worker in paying for them via a payroll deduction program. Second, the employer simply "recommended" that their workers wear slip-resistant shoes, which they were to pay for on their own. Third, and least common, was that employers would provide slip-resistant shoes to their workers for free.

It is common to misinterpret OSHA's Personal Protective Equipment (PPE) requirement, which requires employers to provide PPE footwear to employees at no cost. Slip-resistant footwear is not PPE but rather a component of a work shoe. Work shoes are understood to have three unique elements compared to a street shoe. First, work shoes or boots must provide some form of foot or ankle support. Secondly, they must be water-resistant, and third, they must have a slip-resistant outsole. Also, work shoes are only to be worn in the workplace and are not to be worn as street shoes. In the case of street shoes, there is no requirement for a slip-resistant outsole. PPE footwear is different in that the primary function of PPE footwear is to protect the person's foot. Steel-toed shoes or metatarsal-protective boots are examples of PPE footwear, and OSHA does require employers to provide such footwear to their workers at no cost. This is also true for what OSHA calls "Specialty Footwear," where workers are exposed to specific workplace hazards like excessive heat, water, or other potential conditions that would require them to wear specialized footwear. I then asked them, "So what is the case about?" They hesitated and concluded the phone interview by saying that they will confer with their client about retaining me and will be in touch. It was about two weeks later that they called to inform me that they and their client conferred and agreed to retain me on the case.

THE LAWSUIT

The defendant in the case was the Cargill Company, and I was retained by them in a matter involving slip-resistant work shoes. What had happened was that Cargill had been cited by OSHA in seven of their meatpacking plants in six different states. Each of the citations read exactly the same, and they were all issued on the exact same day. That's odd. How is it that seven different OSHA inspectors all come to author identical citations and issue them on the same day? Clearly something wasn't right here.

What Cargill was being cited and fined for was OSHA's claim that because Cargill required their workers to wear slip-resistant footwear, they had the responsibility to pay for them, which Cargill had not. Cargill did in fact provide their employees who worked in wet areas with specialized rubber knee boots at no cost but did not for their workers' conventional work shoes.

HOW DID IT END?

The case went to trial, whereby I was summoned to appear before the Honorable James R. Rucker Jr. in the matter of "The Secretary of Labor, Complainant vs. Cargill Meat Solutions Corp., Respondent, and UFCW Local No. 2., Authorized Employee Representative." Appearances included two attorneys from the U.S. Department of Labor, Office of the Solicitor for the Complainant; two attorneys from the Denver, Colorado, defense firm who retained me; and finally present was the director of the Occupational Safety and Health Office, United Food and Commercial Workers International Union, Washington, D.C., representing the Authorized Employee Representative.

Given my background and experience in working with OSHA, the Department of Labor attorneys did not challenge my expertise under Daubert (see appendix), and my testimony would be accepted by the court.

On the day before I was called to testify, Cargill's attorney asked the OSHA inspector who issued one of the citations what constituted a "safety shoe." He responded, "I know them when I see them," but could not provide any detail. So when asked what's the difference between a slip-resistant sole and an oil-resistant sole, which are often seen on the bottom soles of work shoes, he didn't know. It was clear that OSHA was in a bind.

My testimony was simple. I defined the differences between a work shoe or boot, a safety shoe or boot, PPE, and the terms slip-resistant and oil-resistant as used on work shoes. OSHA's attorney asked a series of questions related to the interpretation of OSHA's published rules, which concluded by their asking: "Mr. Kendzior, are you telling this court that you know more about OSHA's rules than OSHA does?" I responded, "Well, as it relates to this case, yes!" I was confident that I did in fact know more about OSHA's rules and how they are interpreted and applied than the OSHA officials who were involved in this case, and the Department of Labor's attorney knew it. The judge ordered that the violations by OSHA against Cargill be vacated (see appendix).

What is important to note is the impact this verdict had not only on Cargill but all employers who required their workers to wear slip-resistant footwear. Had Cargill lost, companies like McDonald's, Burger King, and Taco Bueno, to name a few, would have to provide their workers with free shoes. Imagine the cost impact this would have on the fast food industry. Also imagine the potential litigation nightmare that would develop as a result of employers who might be held liable for injuries caused to their workers by the shoes they provided them. In short, had Cargill lost, employers across the country would be in the shoe business and potentially liable for injuries caused by the shoes they made their workers wear.

One big winner would have been the footwear industry. Companies like my previous partner Walmart and others who sell slip-resistant shoes would have seen a sales windfall. OSHA would have set a legal precedent mandating that employers were now to provide their workers slip-resistant shoes at no cost. Cargill and small and large employers won the day; however, I continue to question whether or not this is good for worker safety. I have since concluded that if OSHA wants to mandate slip-resistant footwear for workers across the country, they should do so according to the appropriate regulatory process and not the courts. Since 1992 OSHA has been working on revising their walking and working surface requirements, which I hope will address this issue.

51

DEADLY SIDEWALK!

It was January 14, 2013, when Mr. Roosevelt went for his weekly cancer treatment at the Cancer Care treatment center in Islip, New York. As Mr. Roosevelt was walking on the front entrance sidewalk at a normal pace, he tripped and fell on a raised section of the sidewalk, causing a severe head injury that led to his death. The raised section of sidewalk in question had been previously ground as to bevel the two adjoining sections of the sidewalk. The change in elevation was not marked or painted.

OPINIONS

Production documents revealed that on December 12, 2011, February 27, 2012, and October 10, 2012, the defendant contracted with a concrete grinding company by the name of Wally's Walkway Grinding Company, Inc., for the purpose of grinding down raised/elevated sections of their exterior sidewalks. On August 7, 2008, Wally's Walkway Grinding Company, Inc., was hired by the defendant to raise sunken portions of their exterior sidewalks. After Mr. Roosevelt's trip-and-fall event the defendant contracted with a second company by the name of Uncle Bob's Concrete Services, Inc., for the purpose of removing and replacing sections of their exterior concrete sidewalks.

Based on the evidence provided to me at that time, it was my opinion that the defendant had a long history of their exterior sidewalks becoming raised or elevated and made numerous attempts to resolve the problem. The section of the sidewalk that Mr. Roosevelt tripped on was one that had been previously ground as to bevel the two adjoining sections and create a level surface. However, the previous surface grinding failed to create a truly level surface, and because of such, presented an unreasonably dangerous condition. The sidewalk where Mr. Roosevelt tripped and fell presented an elevated risk of a trip-and-fall event and should have been repaired to create a level walking surface.

Given that many of the defendant's invited guests were their patients, they therefore had the responsibility to provide safe walkways as to accommodate individuals who may have physical limitations or medical impairments.

The standard of care as it pertains to exterior walkways like that in question is promulgated from several nationally recognized consensus standards organizations, including the American Society of Testing and Materials (ASTM) and the American National Standards Institute (ANSI). Exterior walkway safety standards are also codified by way of law via the Americans with Disabilities Act (ADA) and the Texas Accessibility Standard (TAS). It is my opinion that the damaged walkway in question was in direct violation of several of these nationally recognized construction and safety standards, including the ASTM F-1637-10 Standard Practice for Safe Walking Surfaces and the ANSI A1264.2 standard.

Section 5.1.1 entitled "Walkway Surfaces" of the ASTM F-1637-10 standard requires that "walkways shall be stable, planar, flush, and even to the extent possible. Where walkways cannot be made flush and even, they shall conform to the requirements of 5.2 and 5.3."

Section 5.2.1 requires that "Adjoining walkway surfaces shall be made flush and fair, whenever possible and for new construction and existing facilities to the extent practicable." Section 5.7.1 entitled "Exterior Walkway" requires that "Exterior walkways shall be maintained so as to provide safe walking conditions."

Section 5.7.1.2 requires that "Exterior walkway conditions that may be considered substandard and in need of repair include conditions in which the pavement is broken, depressed, raised, undermined, slippery, uneven, or cracked to the extent that pieces may be readily removed."

Section 5.7.2 requires that "Exterior walkways shall be repaired or replaced where there is an abrupt variation in elevation between surfaces. Vertical displacements in exterior walkways shall be transitioned in accordance with 5.2."

Section 301 "General" of the 2010 edition of the ADA requires that "Floor and ground surfaces shall be stable, firm, slip-resistant, and shall comply with 302."

Section 4.5.1 "General" of the Texas Accessibility Standard (TAS) requires that "Ground and floor surfaces along accessible routes and in accessible rooms and spaces including floors, walks, ramps, stairs, and curb ramps, shall be stable, firm, slip-resistant, and shall comply with 4.5. Soft or loose materials such as sand, gravel, bark, mulch or wood chips are not suitable. Cobblestone and other irregular surfaces having a texture that constitutes an obstacle or hazard, such as improperly laid flagstone, shall not be a part of accessible routes, spaces and elements."

Section 4.5.2 "Changes in Level" further requires that "Changes in level up to 1/4 in (6 mm) may be vertical and without edge treatment (see Fig. 7(c)). Changes in level between 1/4 in and 1/2 in (6 mm and 13 mm) shall be beveled with a slope no greater than 1:2 (see Fig. 7(d)). Changes in level greater than 1/2 in (13 mm) shall be accomplished by means of a ramp that complies with 4.7 or 4.8."

It is my opinion that the defendant's management should have properly replaced the elevated section of sidewalk with a new section rather than attempting to "Band-Aid" it by way of edge grinding. Although edge grinding may work as a short-term solution, it does not address the underlying reason for why the concrete "heaved" or became elevated. The defendant should have removed the raised/elevated sections of their exterior sidewalks and replaced them with new concrete to create a level surface.

Based on this information, it is clear that the defendant failed in their duty to protect their invited guests from unnecessary risk of a trip and fall and were negligent in their efforts to protect Mr. Roosevelt. Had the defendant addressed the above-mentioned pedestrian walkway trip hazard prior to Mr. Roosevelt's visit, it is unlikely that he would have tripped and fallen and in turn injured himself.

HOW DID IT END?

Mr. Roosevelt's estate dropped the case against the cancer treatment center.

52

TRIPPED AT THE TRIPMART!

On January 29, 2012, the plaintiff, Melinda, was a thirty-year-old woman who went to her local TripMart convenience store to purchase a breakfast bar. Melinda was eight months pregnant and careful not to go too long without eating. Melinda made her purchase and began to exit the store when she tripped on an exposed metal prong that had extended from the front of a metal A-frame sign on the front sidewalk, directly adjacent to the store's doorway and directly in the path of Melinda's travel. Melinda can be seen on store surveillance video catching her lower leg on the exposed metal prong, causing her to fall forward and in turn injure herself and possibly her unborn child.

INDUSTRY STANDARDS

It is the retail store industry standard to provide for the safety and well-being of their invited guests, which includes the safe use, construction, maintenance, and inspection of retail displays. Such standards are promulgated via authoritative literature including that published by national trade and safety organizations as well as retail underwriting insurance companies. According to the National Safety Council: "Workers must construct displays to prevent breakables and other articles from falling and injuring or tripping the customers. Repair and smooth sharp or broken displays and counters so they do not cut or catch passerby[s]." According to the Great American Insurance Group Loss Prevention

Safety Topics "Retail Shelf Stocking Safety and Fall Safety" publication, "a floor display should not block the aisle, equipped with a protruding base or unstable base." This message is echoed by the Zenith Insurance Company's publication entitled "Management's Role in Slip, Trip and Fall Prevention," which defines a wide range of trip-and-fall contributing factors, including "obstructions." According to Zenith Insurance Company, pedestrian walkway hazards or obstructions "are items that protrude into the normal walking path, such as extension cords, hoses, product storage, material handling equipment guards, concrete posts, parts of equipment, parking lot bumpers, speed bumps and temporary storage/holding areas."

The Zenith publication concludes by stating that "understanding the value of STF prevention and communicating it throughout the organization is the responsibility of management. By incorporating slip, trip and fall awareness into the safety culture of your business, you can help ensure the safety of employees, contractors, visitors and the public. By preventing an incident, you can potentially improve your bottom line." The Zenith Insurance Company, like others who insure the convenience store industry, provides a wide range of resources aimed at reducing pedestrian hazards. Another such publication is that entitled "Retail Safety Orientation Manual." Section F of the manual is entitled "Preventing slips, trips and falls" and encourages retailers like the defendant to "Look out for damaged fixtures and displays. Report to Store Management if any repairs are needed."

OPINIONS

It was my opinion that multiple failures had occurred that could have prevented Melinda's trip-and-fall event and subsequent injuries. First, it is clear based on the photographs of the front entryway that the condition and placement of the sign in question posed a significant and serious trip hazard. Based on the deposition testimony of all the Trip-Mart store employees, they agreed on the condition of the sign whereby the front attachment prong was disconnected from the connecting hole and not properly attached, and therefore they all agreed that the sign was improperly installed. Given the placement of the sign immediately adjacent to the front door and directly in the path of pedestrian travel, the exposed rod posed a potential trip hazard.

Although the store employees admitted that the sign in question was improperly assembled, none of them considered the prong that extended from the sign a trip hazard, and they made no effort to repair the sign in advance of Melinda's visit. Based on the deposition testimony of the defendant's employees, it is clear that they had not received the proper level of training as it relates to identifying pedestrian safety hazards. Walker was the defendant's store supervisor who oversaw nineteen TripMart stores, including the store in question, and was responsible for the operation of the stores within his control, which included ensuring that store managers and employees were in compliance with "all TripMart's policies, procedures and standards," which included store safety. However, when he was asked if he had ever interacted with TripMart's safety department, he replied, "No." When it came to the safe use of equipment, he was only aware of the store safety facility walk, which under the section entitled "Customer Safety," subsection "General Housekeeping and Safety—Outside," states that employees are to ensure that "Customers can safely move around displays" and to "Make sure store facilities are safe for customers as well as store team." Although the defendant's employees were responsible for customer and employee safety, they were not properly trained and failed to correct a serious pedestrian trip hazard: the exposed metal display prong in question. In short, the defendant's employees were responsible for the safety of their customers. The lack of discipline by the employees at the store after Melinda's incident indicates a lack of proper supervision by the defendant's management. When safety rules are not followed, management needs to retrain employees to ensure that they understand what is an unsafe condition and, if necessary, reprimand workers who continue to violate the company's policies. This opinion is further supported by the store policy entitled "Trips and Falls," which only has a total of three bullet points. None make any mention of how to identify or prevent any type of walkway slip or trip hazard other than posting wet-floor signs and encouraging employees to "Pay attention to where you are going." Needless to say, the defendant's safety and training materials and safety policies are grossly incomplete and not in compliance with the industry standard.

It is my conclusion that Melinda's trip-and-fall event could have been prevented had the defendant (a) connected the A-frame sign's metal prong properly as to not be exposed from the sign, (b) frequently inspected the sign in question to identify any defects/safety hazards such as the exposed prong, (c) received proper safety training specifically as it relates to proper use, inspection, and maintenance of their pedestrian walkway and the retail signs they place at their front entrances, and (d) TripMart's management provided adequate supervision and enforcement of the employees as it relates to their safety practices.

Also, it is my view that the defendant knew or should have known that the metal A-frame displays that they placed on their front sidewalk directly adjacent to their front doors may from time to time present a potential pedestrian trip hazard, which would therefore require timely inspections; according to store employees, these were not performed with any frequency.

In summary, it is my conclusion that the defendant failed to exercise reasonable care as it relates to the proper installation, maintenance, and inspection of the metal sign in question and because of such was the proximate cause of Melinda's trip-and-fall event. The defendant further failed to provide proper training and supervision. Because the injury to Melinda was due to a system failure and not an isolated incident, such injuries are bound to continue to occur until such time that the defendant takes safety seriously and performs adequate training and supervision to ensure that their store-level employees understand how important safety is to the defendant's customers.

HOW DID IT END?

The case went to trial, where the jury rendered a defense verdict. After her fall Melinda was placed in the hospital for two weeks. Shortly thereafter she gave birth to a beautiful baby boy she named Trip.

53

STICK A FORK IN HER!

It was 4:00 p.m. on June 12, 2013, when Kathryn and her husband, Scott, went to the Carpet Palace located in Alexandria, Louisiana, to purchase carpeting for their home. While in the showroom, Kathryn was instructed by the store's salesperson, Randy, to accompany him into their warehouse to look at carpet they had in stock. As Kathryn was walking through the defendant's warehouse, talking to the salesperson that was drawing her attention toward carpet samples he had in his hand, she tripped and fell over a forklift steel fork that was elevated approximately six inches above ground level, causing her to fall onto the concrete warehouse floor, inflicting serious injury to her body. After tripping and falling Kathryn was helped to her feet by Randy along with two of his employees. Kathryn was told that one of the warehouse workers had accidentally left the forklift in this location and knew that customers would be coming in and out of the warehouse.

INDUSTRY STANDARDS AND CODES

Section 7.1.10 "Means of Egress" of the National Fire Protection Association (NFPA) Life Safety Code 101 requires that "Means of egress shall be continuously maintained free of all obstructions or impediments to full instant use in the case of fire or other emergency." The forklift as parked at the time of Kathryn's trip-and-fall event was in the means of egress and in violation of this section of the NFPA Life Safety Code 101.

The American National Standards Institute (ANSI) A1264.2-2006 standard describes the procedures by which pedestrians are to be safeguarded against and properly warned as to impending walkway slip-and-trip hazards. Listed below are applicable excerpts from the ANSI A1264.2-2006 standard.

Section 9 "Warnings/Barricades," Subsection 9.1 "General" requires that "A warning shall be provided whenever a slip or trip hazard has been identified until appropriate corrections can be made, or the area barricaded." Section E9.1 states that "Slip/trip hazards should be eliminated by design, and arrangement if possible. The next priority is to guard the hazard if possible, and the last priority is to warn of the hazard. The intent is to eliminate and/or reduce, as much as possible, the potential for injury."

Section 9.1.1 "Alternate Route" requires that "When there is a slip/fall hazard, which covers an entire walkway, making it difficult to safely route personnel around the hazard, barricades shall be used to limit access (see Section 10.1). If appropriate, assign a person(s) to detour pedestrians, in conjunction with the appropriate use of warning signs until the barricade can be erected or the hazard removed." Section 9.2 "Signage" states that "The signage reference for warning signs used for slip/fall hazards shall be ANSI Z535.2, Standard for Environmental and Facilities Safety Signs."

Section 9.4 "Placement" requires that "Warning signs shall be placed at approaches to, or around, areas where slip/fall hazards exist. These devices shall surround or be placed around the perimeter of the hazardous area to clearly demarcate the location of the potential hazard."

Section 9.4.1 "Unmitigated Hazards" states that "In cases where hazards cannot be mitigated, warning signs and barricades shall be used to reroute traffic." Section 9.4.2 "Removal of Hazard" requires that "When the hazard has been eliminated or controlled, the sign and/or barricade shall be promptly removed."

OPINIONS

Although it is customary for flooring retailers to provide guest access to their warehouse for the purpose of showing their inventory, retailers must provide for their invited guests' safety. The defendant's forklift in question was improperly parked, with forks raised as to be in the direct path of pedestrian travel; the defendant therefore failed to provide adequate protection to prevent their invited guests from catching their feet on the exposed and elevated steel forklift fork(s).

It is the industry standard to store all material-handling equipment such as forklifts, pallet jacks, and so on, in a location where customers would not be exposed. Warehouse equipment like the forklift in question may pose a significant trip hazard if their forks are exposed to pedestrian travel. Forks that are elevated above ground level by six inches or more pose a serious potential trip hazard. Flooring retailers like the defendant are aware that their customers are not familiar with the means and methods of transporting and storing roll carpeting and/ or flooring materials in general, and therefore, when invited into the warehouse, the defendant has the responsibility to provide for their safety, specifically as it relates to their walkways. Material-handling equipment that is stored as to be in the pedestrian path of travel can lead to a guest injury.

Furthermore, there are a series of warehouse safety controls that directly affect how flooring retailers design the layout of their warehouse/showroom to accommodate shoppers. Given that the defendant's warehouse functioned both as a place of storage as well as an extension of their retail showroom, the rules that govern walkway safety must ensure the safety of both their workers, who operate a range of powered equipment, as well as members of the public, who are not familiar with warehouse operations. Such controls should include procedures and training on the proper storage of warehouse equipment and supplies, safety inspections by trained employees, as well as supervision by management personnel to ensure invited guests are not exposed to any industrial work hazards.

The defendant chose to locate their warehouse forklift directly within the path of travel, whereby Kathryn was directed to and forced to walk around the forklift, whose forks were elevated, thus increasing the risk that she could trip over them as she was looking at rolls of carpeting. The risk of an accidental foot contact with the elevated forks was further enhanced by the fact that Kathryn's attention was directed by the defendant's salesperson toward the store's rolls of carpeting and not down toward the floor or the forklift. Also, the steel forks in question were of a color similar to that of the concrete floor beneath them and blended in with the surrounding area as to camouflage them; this in turn elevated the risk of a pedestrian trip-and-fall event.

It is my opinion that Kathryn's trip-and-fall event could have been prevented had the defendant (a) not parked the forklift in an area where shoppers would be expected to walk, (b) barricaded the area where the forklift was parked to restrict access, (c) provided a verbal warning cautioning shoppers to stay clear from the exposed forks, and (d) provided a visual warning sign cautioning shoppers to stay clear from the exposed forks.

It is my conclusion that the defendant knew or should have known about proper storage of their warehouse equipment as to prevent a guest trip hazard. It is my opinion that the defendant failed to do such and in turn failed to exercise reasonable care as it relates to the forklift in question; therefore they were directly responsible for Kathryn's trip-and-fall event and subsequent injuries.

HOW DID IT END?

It hasn't ended yet. This case is still an open file.

54

WHEEL STOPPER!

It was a few days after Linda's fiftieth birthday when she went to get her nails done at a local Southlake, Texas, nail salon, which she received as a gift from her husband, Reed. Shortly after parking her car in the nail salon's parking lot, Linda exited her car and began walking toward the front entrance. After just a few steps Linda tripped and fell on a concrete wheel stop that was placed between the parking spaces, causing her to fracture her hand and wrist. The wheel stop in question was not painted in a contrasting color (safety yellow), and the parking lot was dimly lit.

INDUSTRY STANDARDS

The need for safe walkway surfaces is outlined in several industry standards including the American Society of Testing and Materials (ASTM) F-1637-10 "Standard Practice for Safe Walking Surfaces," which establishes the standard of care for wheel stops. Section 4.2 of the F-1637-10 standard entitled "Walkway Changes in Level" requires the following: Section 4.7.1 requires that "Exterior walkways shall be maintained so as to provide safe walking conditions."

Section 4.7.1.2 requires that "Exterior walkway conditions that may be considered substandard and in need of repair include conditions in which the pavement is broken, depressed, raised, undermined, slippery, uneven, or cracked to the extent that pieces may be readily removed."

The standard also addresses the subject of proper illumination and use of wheel stops. Section 5.5.1 requires that "Minimum walkway illumination shall be governed by the requirements of local codes and ordinances or, in their absence, by the recommendations set forth by the Illuminating Engineering Society of North America (IES) (Application and Reference Volumes)." Section 5.5.2 requires that "Illumination shall be designed to be glare free." Section 5.5.3 requires that "Illumination shall be designed to avoid casting of obscuring shadows on walkways, including shadows on stairs that may be cast by users." Section 5.5.4 requires that "Interior and exterior pedestrian use areas, including parking lots, shall be properly illuminated during periods when pedestrians may be present."

Section 9.1 states that "Parking lots should be designed to avoid the use of wheel stops." Section 9.2 requires that "Wheel stops shall not be placed in pedestrian walkways or foreseeable pedestrian paths." Section 9.3 requires that "Wheel stops shall be in contrast with their surroundings." Section 9.4 specifies that "Wheel stops shall be no longer than 6 ft. (1.83 m) and shall be placed in the center of parking stalls. The minimum width of pedestrian passage between wheel stops shall be 3 ft. (0.91 m)." Section 11.2 states that "When relying on applications of color as a warning, provide colors and patterns that provide conspicuous markings for the conditions being delineated, their surroundings, and the environment in which they will be viewed by users. Bright yellow is a commonly used color for alerting users of the presence of certain walkway conditions. When properly applied and maintained, other colors can also provide effective warnings."

OPINIONS

It was my opinion that the wheel stop in question was not in compliance with the industry standard of care for pedestrian safety in parking lots, and therefore it is my opinion that the defendant failed to provide a reasonably safe walking surface for their invited guests. Furthermore, because of the inherent danger associated with wheel stops and because the defendant's customers had to step over the wheel stop in order to access the building's sidewalk and front entry, it is my opinion that the parking lot in question should have been designed as not to use a wheel stop. Wheel stops increase the risk of a pedestrian trip-and-fall event, and when improperly placed, like in this case, elevate the risk even higher.

Since the wheel stop was placed on the defendant's property by the defendant and therefore was under their control, they were therefore aware of its location and the potential trip hazard it represented. The defendant was aware of such condition prior to Linda's visit and therefore had a responsibility to (a) remove the wheel stop, and/or (b) relocate the wheel stop as to be centered in the parking space, and (c) warn their invited guests as to the presence of the known trip hazard via proper hazard identification. The defendant failed to comply with any of the suggested remedies but rather opted to do nothing and failed to correct the wheel-stop trip hazard, which in turn was in my opinion the proximate cause of Linda's trip-and-fall event and related injuries.

It is my conclusion that the improperly placed wheel stop in question posed an unreasonable risk of harm to anyone parking in the defendant's parking lot. Since the safety of the parking lot and wheel stop was within the control of the defendant, it is my opinion that the defendant failed in their duty to protect the public, specifically Linda, from unnecessary risk of injury and therefore were directly responsible for her injuries.

HOW DID IT END?

The nail salon settled with Linda for $92,000. Her wrist and arm healed but with it came an inflammation of arthritis.

55

CRAZY BUFFALO WINGS!

At 12:45 p.m. on February 19, 2013, Wanda and her son Gregory went to their local Crazy Buffalo Wings restaurant located in Albuquerque, New Mexico, to eat lunch. After eating, Wanda and Gregory began to exit the restaurant when Wanda tripped and fell over a carpet floor mat located at the restaurant's interior entrance. Wanda stated that she did not notice the mat; however, after her fall she saw a "hump" in the mat, which she believed was the cause of her trip and fall. At the time of her trip-and-fall event Wanda was wearing a pair of dark-colored "loafer-type" street shoes that were in good condition.

In her deposition Wanda said that she saw a "hump in the mat" that caused her foot to go under the mat, causing her to trip and fall. Wanda also stated that "I know my shoe went under it, because the rug rolled up under me" and that "I didn't push it up. It just went under it." Once her foot engaged, "It (the mat) went up to her ankle."

Gregory stated in his deposition that he did not actually witness his mother's trip-and-fall event but soon afterward saw that she was on the floor with the mat "rolled up under her" and that "right after she had fallen, or right after Gail the restaurant manager straightened it up, because when she was still down on the mat, it was rolled up." He further stated that he witnessed that Gail "straightened up the mat," whereby it "fell back with a hump." Shortly thereafter Gregory stated that Gail "folded it up and took it away."

Gail stated that her only training in safety occurred during the initial eight weeks of training that she received when she first joined the company. As the manager, she is responsible for designating areas for mats, and she placed the mat in question at its location. She also stated that she inspected the mat in question on the day it was delivered but that she had not received any training as it relates to mat inspection other than "common sense." She stated that after Wanda had tripped and fallen she had inspected the mat and found it to be in "good condition" and that she did not see any "bumps or rolls" in the mat. After Wanda's fall Gail did not take any photographs of the scene or of the mat in question, and she did not review the surveillance camera video, which may have captured the event. Gail then "pulled the mat" in question from its location in the lobby and put it in the back of the restaurant for the rental mat supplier to pick up. Finally, Gail said that she prepared an incident report from the information provided to her from Wanda.

James was the restaurant's general manager, and he stated that neither he nor Gail "pulled the mat" after Wanda fell; it remained at the front lobby the rest of the day. He further stated that the "front-of-the-house inspections" are done three times a day; however, mat inspections are not part of the inspection procedure.

INDUSTRY STANDARDS

The American Society of Testing and Materials (ASTM) F-1637-10 standard entitled "Standard Practice for Safe Walking Surfaces" establishes the standard of care by which property owners shall use, maintain, and inspect entranceway floor mats like that used by the defendant. Listed below are sections from the standard and two additional nationally recognized consensus standards that apply to this case:

Section 5.3.2 states that "Carpet on floor surfaces shall be routinely inspected." Subsection 5.4.5 of the ASTM F-1637-10 "Standard Practice for Safe Walking Surfaces" states that "Mats, runners and area rugs shall be provided with safe transition from adjacent surfaces and shall be fixed in place or provided with slip resistant backing."

A second industry standard was also in violation. Section 7 entitled "Mats and Runners" of the American National Standards Institute (ANSI) A1264.2-2012 states in Section 7.3 that "Mats and runners shall be installed and maintained not to move when in use." Subsection 7.4:

"Mats shall be installed so that they do not create tripping hazards." Subsection 7.5: "Mats and runners shall be routinely inspected and adequately maintained to identify and correct conditions such as buckling, edge cutting and other defects. Damaged mats shall be promptly replaced." Section 8 "Reduction of Hazards Related to Matting" of the ANSI/NFSI B101.6-2012 "B101.6 Standard Guide for Commercial Entrance Matting in Reducing Slips, Trips and Falls" requires that "Facility Management shall take precautions to avoid mats becoming a hazard in and of themselves through regular inspection, particularly during inclement weather." Subsection 8.2 "Rippled, Curled or Torn Edges" requires that "When mats ripple, curl or have torn edges the mats shall be removed from service and replaced with mats that lay flat." E8.2 states that "Sometimes rippled edges can't be made to lay flat with tape due to permanent damage. In this circumstance, the mats shall be replaced." Subsection 8.3 "Buckling" states that "If a mat buckles, either the condition that caused the mats to buckle shall be corrected or the mats shall be secured or removed from service and replaced." E8.3 further states that "To increase the dimensional stability and reduce the tendency of matting to buckle, the user should consider purchasing heavier mats such as mats with thicker backing."

Given that there were no eyewitnesses who actually saw Wanda trip and fall, the following opinions are based on the existing facts in evidence. Having reviewed the facts and testimony of the interested parties, it is my opinion that Wanda's description of her trip-and-fall event represents a fair and accurate cause-and-effect relationship. Wanda's description and subsequent injuries support the fact that she did in fact engage the floor mat in question in such a way as to cause her to trip and fall. Wanda's body position, along with the floor mat's condition, are consistent with a mat-related trip-and-fall event.

Carpet mats like that used by the defendant are common in the restaurant industry; however, the location of the mat in question appeared to serve no purpose. Entranceway floor matting is used primarily to remove both dry and wet materials from incoming pedestrians' shoes to keep the facilities' floors clean and dry. The location of the mat in question was more than ten feet from the front doors and at such a distance would not provide for the removal of soil or moisture. Given that Gail stated that she was the person responsible for mat placement, it is my opinion that she made a poor choice in placing the mat where she did. Furthermore, because of pedestrian foot traffic, entranceway floor matting is prone to movement and buckling and therefore requires frequent inspections as to ensure that it remains in place and does not create a trip hazard.

It is clear, having read Gail's deposition testimony, that her recollection of Wanda's trip-and-fall event differed significantly. Gail authored the defendant's guest incident report dated February 19, 2013, and stated that "the rug pulled up as she stepped into it" and that "Wanda stepped towards the rug, kicked it up, and tripped over it." Gail also claimed that such information was provided by Wanda; however, Wanda did not author, review, or sign the incident report and testified in her deposition that such information as recorded by Gail was an inaccurate description of her trip-and-fall event. A second contradiction is the claim by Wanda and her son Gregory that after her fall Gail removed the mat from the lobby—a claim that Gail agreed with but was contradicted by James in his deposition. James stated that the floor mat was not removed and that it remained in place the remainder of the day. One can only assume that if Gail did in fact remove the mat in question, it was because she perceived it as having some type of defect that was related to Wanda's fall.

Based on the facts as presented to me, it is my opinion that Wanda was exercising reasonable care as she entered and exited the defendant's restaurant and egressed along her path of travel at a normal pace. Given such, it is my opinion that the entranceway floor mat in question presented an unreasonable risk of harm to individuals egressing across it and was not in compliance with the nationally recognized safety standards listed above. Failure to observe such standards suggests that the defendant failed to provide reasonable care as it relates to the safe use, inspection, and care of their entranceway mat.

It is my opinion that had the floor mat in question been installed as to lie flat as required by the above-named industry standards, Wanda would not have tripped and fallen. The defendant should have provided for more frequent inspections of the mat in question, which according to Gail was only done once a day. Had the defendant inspected the mat in question more frequently, they would have noticed that it was not flat and then had the opportunity to flatten it or remove it from service. Furthermore, the defendant could have secured the mat in question as to mitigate its movement/buckling. Such failure to properly secure the floor mat in question and failure to recognize and replace the mat given its condition and the hazard it presented, were in my opinion, the direct cause of Wanda's trip and fall and related injuries.

Had the defendant addressed the above-mentioned walkway hazard prior to Wanda's visit, and had they taken proper care of the mat, which was within their control, it is unlikely that Wanda would have tripped and fallen and subsequently injured herself. It is therefore my conclusion that the defendant failed to provide for the reasonable care of their entranceway floor mat and in turn created an unreasonably dangerous condition.

HOW DID IT END?

Wanda and the restaurant settled their case for $140,000.

APPENDIX

US 20120094057A1

(19) **United States**

(12) **Patent Application Publication**   (10) Pub. No.: **US 2012/0094057 A1**
Bartlett   (43) Pub. Date: **Apr. 19, 2012**

(54) POROUS ANTI-SLIP FLOOR COVERING

(76) Inventor: Joel Patrick Bartlett, Fort
                   Lauderdale, FL (US)

(21) Appl. No.:   13/228,525

(22) Filed: Sep. 9, 2011

Related U.S. Application Data

(60) Provisional application No. 61/393,012, filed on Oct.
     14, 2010, provisional application No. 61/393,503,
     filed on Oct. 15, 2010.

Publication Classification

(51) Int. Cl.
     D04H 11/00   (2006.01)
     B32B 37/14   (2006.01)

(52) U.S. Cl. 428/92; 156/62.2

(57)   **ABSTRACT**

A flooring assembly and method of manufacturing a flooring
assembly includes forming a. working surface by depositing
a plurality of heated flexible strands of material onto a surface
and cooling the plurality of heated flexible strands so that a
first end of each strand is directly physically coupled to a first
end of at least one other strand in the plurality of strands so
that the coupled first ends of the plurality of strands define a
plurality of openings and a second end, opposite the first end,
of each strand is physically independent from each second
end of each of the strands in the plurality of strands so that
each second end of each strand is able to move independent of
the second end of each other strand. The method further
includes providing a mesh underlayment having an upper
surface, an opposing lower surface, and defining a plurality of
openings spanning from the upper surface to the lower surface and permanently adhering the coupled first ends of the
plurality of strands to the upper surface of the underlayment.

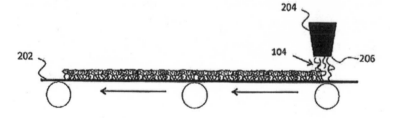

OCCUPATIONAL SAFETY AND HEALTH REVIEW COMMISSION
721 19TH Street, Room 407
Denver, CO 80202-2517

SECRETARY OF LABOR,

Complainant,

v.   OSHRC DOCKET NO. 10-1156

CARGILL MEAT SOLUTIONS CORP.,

Respondent,

and

UFCW Local No. 2,

Authorized Employee Representative.

APPEARANCES:

Andrea C. Luby, Esquire, Susan J. Willer, Esquire
U.S. Department of Labor, Office of the Solicitor, Kansas City, Missouri
For the Complainant.

Rodney L. Smith, Esquire, Patrick J. Miller, Esquire
Sherman & Howard, LLC, Denver, Colorado
For the Respondent.

Jacqueline Nowell, Director, Occupational Safety and Health Office
United Food and Commercial Workers International Union, Washington, D.C.
For the Authorized Employee Representative.

BEFORE: James R. Rucker, Jr.
             Administrative Law Judge

DECISION AND ORDER

This proceeding is before the Occupational Safety and Health Review Commission ("the Commission") under section 10(c) of the Occupational Safety and Health Act of 1970, 29 U.S.C. § 651 *et seq.* ("the Act"). On January 13, 2010, the Occupational Safety and Health

United States
CONSUMER PRODUCT SAFETY COMMISSION

Search CPSC 🔍

HOME   RECALLS   SAFETY   REGULATIONS,   RESEARCH &   BUSINESS &   NEWSROOM   ABOUT CPSC
                 EDUCATION   LAWS &   STATISTICS   MANUFACTURING
                               STANDARDS

Home » Recalls » Tristar Products Recalls AquaRug Shower Rugs » Tristar Products Recalls AquaRug Shower Rugs

Tristar Products Recalls AquaRug Shower Rugs Due to Fall Hazard

 En Español

f 🐦 ✉ 🖶 ➕   322

Recall date: January 28, 2016   **Recall number: 16-084**

1 of 2 photos   « Previous   Next »

Beige AquaRug Four Suction Cup Shower
Rug for Bathtubs

Recall Summary

Name of product:
AquaRug shower rugs

Hazard:
The four suction cups on the underside of the rugs can fail to prevent slipping, posing a fall hazard to the user.

Remedy:
🔄 Replace

Consumer Contact:
Tristar Products toll-free at 888-770-7125 from 7 a.m. to 6 p.m. CT Monday through Friday, or visit the firm's website at www.tristarproductsinc.com and click on "Aqua Rug Recall" for more information.

Recall Details

Report an Incident Involving this Product

Units:
About 1.4 million (in addition 70,000 rugs were sold in Canada)

Description:
This recall involves Aqua Rugs with four plastic suction cups. The rugs are intended to provide a slip-resistant surface in the shower or bathtub. The rugs were sold in beige and clear, and in two sizes: 29.5 inches by 17.25 inches for use in the bathtub, and 21.75 inches by 19.75 inches for use in a shower stall. The rugs have a plastic border and only four plastic suction cups, one affixed to the underside of each corner of the rug. "AquaRug" and "As Seen On TV" are printed on the front of the cardboard packaging.

Incidents/Injuries:
Tristar has received 60 reports of consumers falling in the shower or bathtub while on the recalled four suction cup rugs, including 30 reports of injuries such as bruises, cuts, and fractured or broken bones.

Remedy:
Consumers should immediately stop using the recalled shower rugs and contact Tristar for instructions on how to dispose of the rugs and to obtain a free replacement rug.

Sold At:
Bed Bath & Beyond, Dollar General and other retail stores nationwide, online at Amazon.com and BuyAquaRug.com, by Tristar through direct response television commercials, and through a live television show on QVC, from July 2012 to September 2015 for between $18 and $26.

Importer(s):
Tristar Products Inc., of Fairfield, N.J.

Manufactured In:
China

The U.S. Consumer Product Safety Commission is charged with protecting the public from unreasonable risks of injury or death associated with the use of thousands of types of consumer products under the agency's jurisdiction. Deaths, injuries, and property damage from consumer product incidents cost the nation more than $1 trillion annually. CPSC is committed to protecting consumers and families from products that pose a fire, electrical, chemical or mechanical hazard. CPSC's work to help ensure the safety of consumer products - such as toys, cribs, power tools, cigarette lighters and household chemicals — contributed to a decline in the rate of deaths and injuries associated with consumer products over the past 40 years.

Federal law bars any person from selling products subject to a publicly-announced voluntary recall by a manufacturer or a mandatory recall ordered by the Commission.

To report a dangerous product or a product-related injury go online to www.SaferProducts.gov or call CPSC's Hotline at 800-638-2772 or teletypewriter at 301-595-7054 for the hearing impaired. Consumers can obtain news release and recall information at www.cpsc.gov, on Twitter @USCPSC or by subscribing to CPSC's free e-mail newsletters.

Media Contact
Please use the below phone number for all media requests.

Phone: (301) 504-7908
Spanish: (301) 504-7800

View CPSC contacts for specific areas of expertise

INDEX

ABOUT THE AUTHOR

Russell J. Kendzior is president of Traction Experts, Inc., and founder of the National Floor Safety Institute (www.nfsi.org). Internationally recognized as the leading safety expert in slip, trip, and fall accident prevention, Kendzior has been retained in over 400 slip, trip, and fall lawsuits. He has expertise in cases involving ADA and OSHA violations, and he is a leading researcher, educator, and expert consultant to the manufacturing, insurance, and safety industries. He is also author of *Falls Aren't Funny: America's Multi-Billion Dollar Slip-and-Fall Crisis* (2010) and *Slip-and-Fall Prevention Made Easy: A Comprehensive Guide to Preventing Accidents* (1999).